BIOLOGY OF EMERGING VIRUSES

SARS, Avian and Human Influenza, Metapneumovirus, Nipah, West Nile, and Ross River Virus

ANNALS OF THE NEW YORK ACADEMY OF SCIENCES
Volume 1102

BIOLOGY OF EMERGING VIRUSES
SARS, Avian and Human Influenza, Metapneumovirus, Nipah, West Nile, and Ross River Virus

Edited by Sunil K. Lal

Published by Blackwell Publishing on behalf of the New York Academy of Sciences
Boston, Massachusetts
2007

Library of Congress Cataloging-in-Publication Data

Biology of emerging viruses : avian and human influenza, metapneumovirus, Nipah virus, Ross River virus, and SARS / edited by Sunil Lal.
p. ; cm. – (Annals of the New York Academy of Sciences, ISSN 0077-8923 ; v. 1102)
Includes bibliographical references.
ISBN-13: 978-1-57331-690-3 (alk. paper)
ISBN-10: 1-57331-690-3 (alk. paper)
1. Virus diseases–Congresses. 2. Medical virology–Congresses. 3. Emerging infectious diseases–Congresses. I. Lal, Sunil K. II. New York Academy of Sciences. III. Series.
[DNLM: 1. Communicable Diseases, Emerging–Congresses. 2. Virus Diseases–Congresses. 3. RNA Viruses–genetics–Congresses. 4. Viral Proteins–genetics–Congresses. W1 AN626YL v.1102 2007 / WC 500 B615 2007]

QR201.V55B56 2007
571.9'92–dc22

2007005113

The *Annals of the New York Academy of Sciences* (ISSN: 0077-8923 [print]; ISSN: 1749-6632 [online]) is published 28 times a year on behalf of the New York Academy of Sciences by Blackwell Publishing with offices at 350 Main St., Malden, MA 02148 USA; 9600 Garsington Road, Oxford, OX4 2ZG UK; and 600 North Bridge Rd, #05-01 Parkview Square, 18878 Singapore.

Information for subscribers: For new orders, renewals, sample copy requests, claims, changes of address and all other subscription correspondence please contact the Journals Department at your nearest Blackwell office (address details listed above). UK office phone: +44 (0)1865 778315, fax +44 (0)1865 471775; US office phone: 1-800-835-6770 (toll free US) or 1-781-388-8599; fax: 1-781-388-8232; Asia office phone: +65 6511 8000, fax; +44 (0)1865 471775, Email: customerservices@blackwellpublishing.com

Subscription rates:
Institutional Premium The Americas: $4043 Rest of World: £2246
The Premium institutional price also includes online access to full-text articles from 1997 to present, where available. For other pricing options or more information about online access to Blackwell Publishing journals, including access information and terms and conditions, please visit www.blackwellpublishing. com/nyas
*Customers in Canada should add 6% GST or provide evidence of entitlement to exemption.
**Customer in the UK or EU: add the appropriate rate for VAT EC for non-registered customers in countries where this is applicable. If you are registered for VAT please supply your registration number.

Mailing: The *Annals of the New York Academy of Sciences* is mailed Standard Rate. Mailing to rest of world by DHL Smart & Global Mail. Canadian mail is sent by Canadian publications mail agreement number 40573520. **Postmaster:** Send all address changes to *Annals of the New York Academy of Sciences*, Blackwell Publishing Inc., Journals Subscription Department, 350 Main St., Malden, MA 02148-5020.

Membership information: Members may order copies of *Annals* volumes directly from the Academy by visiting www.nyas.org/annals, emailing membership@nyas.org, faxing 212-298-3650, or calling 800-843-6927 (US only), or 212-298-8640 (International). For more information on becoming a member of the New York Academy of Sciences, please visit www.nyas.org/membership. Claims and inquiries on member orders should be directed to the Academy at email: membership@nyas.org or Tel: 212-298-8640 (International) or 800-843-6927 (US only).

Printed in the USA. Printed on acid-free paper.

Annals are available to subscribers online at the New York Academy of Sciences and also at Blackwell Synergy. Visit www.blackwell-synergy.com or www.annalsnyas.org to search the articles and register for table of contents e-mail alerts. Access to full text and PDF downloads of *Annals* articles are available to nonmembers and subscribers on a pay-per-view basis at www.blackwell-synergy.com and www.annalsnyas.org.

The paper used in this publication meets the minimum requirements of the National Standard for Information Sciences Permanence of Paper for Printed Library Materials, ANSI Z39.48_1984.

ISSN: 0077-8923 (print); 1749-6632 (online)
ISBN-10: 1-57331-690-3 (paper); ISBN-13: 978-1-57331-690-3 (paper)

A catalogue record for this title is available from the British Library.

ANNALS OF THE NEW YORK ACADEMY OF SCIENCES

Volume 1102
April 2007

BIOLOGY OF EMERGING VIRUSES

SARS, Avian and Human Influenza, Metapneumovirus, Nipah, West Nile, and Ross River Virus

Editor
Sunil K. Lal

This volume is the result of a meeting entitled **Emerging Viral Infectious Diseases,** co-sponsored by the American Society for Microbiology and the Vietnam Association for Microbiology and held on April 4–5, 2005 in Hanoi, Vietnam.

CONTENTS

Preface

On April 4–5, 2005, a meeting entitled Emerging Viral Infectious Diseases was held in Hanoi, Vietnam. This meeting was developed as a result of previous interactions and discussions held in Vietnam among the American Society for Microbiology (ASM), Ambassador Sunil Lal, and Vietnam Association for Microbiology (VAM) officials, the latter of which fostered a relationship between the two professional societies. The meeting was cosponsored by the ASM and the VAM; the goal was to provide a common platform for Vietnamese scientists and ASM international members located in the Asia–Pacific region to come together to discuss emerging viral-infectious diseases. Special emphasis was given to the newly emerging viral diseases, like the bird flu virus and the recent severe acute respiratory syndrome (SARS) epidemic that rocked Southeast Asian countries, inasmuch as the global response to the SARS epidemic has demonstrated strengths and weaknesses in national and international capabilities to address all infectious disease challenges.

The ASM, the largest scientific society for one single group of disciplines in the world, includes more than 9,000 international members (roughly 30% of the total membership) from various ethnic, language, and cultural backgrounds, who live on five continents outside the borders of the United States. Worth mentioning, the ASM Global Outreach Program, which currently serves 480 members worldwide, provides free membership and services to microbiologists in the lowest tier of countries, as identified by the UN Human Development Index (33 countries in Africa and Asia, in addition to Haiti are currently eligible). International members constantly inspire the society's international activities. They are also the faces, voices, and strength of the ASM in global microbiology today.

The VAM has been very active within Vietnam, and its members are scattered all over the vertical length of Vietnam. VAM organizes and holds its annual meetings in various parts of the country, with a robust schedule of scientific exchanges among its own members and invited delegates. Currently, VAM is concentrating its efforts on the bird flu problem and is partnering with agricultural and veterinary institutions throughout the country for H5N1 surveillance.

In these times of globalization, the microbiological disciplines have come to play a new, important role in the evolution of many countries. As we write, the emergence or reemergence of infectious diseases has already taken a significant toll on many populations and economies, developing and developed alike, while new scientific advances and tools are around the corner. This will likely be one of the major challenges of the upcoming years. In this context, we

Ann. N.Y. Acad. Sci. 1102: ix–x (2007). © 2007 New York Academy of Sciences.
doi: 10.1196/annals.1408.000

believe that involving scientists in appropriate scientific dialogue and in the development of multilateral policies will be essential.

The ASM International Membership Committee believes that improving the flow of ideas among ASM members and members of other national societies in the form of member networks contributes significantly to this dialogue. The International Ambassadors Program, created to provide services to our international members, has clearly demonstrated here how the access to local knowledge and networks represents a significant service to the local scientific community. On the model of real-life diplomats, our ambassadors already serve as local experts for our society and as liaisons with the other national societies. It is hoped they will play a major role in furthering a global international vision and will be major actors in our common future.

Researchers and scientists in Vietnam have a lot to gain from interaction with other scientists in the region; conversely, there are many overlapping areas of interest, from which international scientists can learn from the experiences of their Vietnamese counterparts. Efforts in liaising between the ASM and the VAM culminated in the organizing of this 2-day joint meeting on Emerging Infectious Diseases. Beyond academic exchange, the large measure of goodwill encountered has already created many opportunities and has opened the doors for future collaborations and networking among participants from various working groups in different parts of Asia. A small step has been made, and although we know that much lies ahead, we are confident this is only the beginning.

—SUNIL K. LAL
International Centre for Genetic Engineering and Biotechnology
New Delhi, India
—BERNARD J. BUCHER
American Society for Microbiology
Washington, D.C.

Foreword

Emerging infectious diseases have captured the world's attention. The SARS outbreak and existing threats of pandemic influenza and bioterrorism have made us realize and be aware of the fact that new infectious diseases may emerge at any time and in any place. Emerging diseases transcend species barriers, both human and animal, and obey no national boundaries.

The boom in livestock production and globalization, with an enormous increase in free trade of animals and food products compounded by modern living conditions and low hygiene in economically weaker regions of the world, ensures that factors responsible for disease emergence are in plenty. Major responsible factors include ecological changes, such as changes due to agriculture or economic development or to anomalies in climates; human demographic changes and behavior; travel and commerce; technology and industry; microbial adaptation and change; and the breakdown of public health measures. Epidemiological studies currently do not predict accurately many emerging new pathogens. Only the evolutionary perspectives provide us rough theoretical estimates for selective processes in pathogen populations. Surveillance and monitoring thus remain the most important methods to recognize and detect early new diseases. In light of the complexity and diversity of likely new emerging diseases, such surveillance should be targeted toward early recognition of disease syndromes rather than at identifying microbial diseases. The complex and rapid developments in international trade coupled with an increased demand for abundant as well as inexpensive food originating from animals or animal products, demand careful attention from both veterinary as well as human health officials. There exists a huge potential for the study of zoonosis among related professionals as we understand the science of emerging diseases, especially under the current observation that among the 60 viral species reported to be associated with bats, 59 are RNA viruses, which are potentially important in the generation of emerging and reemerging infections in humans. The prime examples of these are the lyssaviruses and Henipavirus. The transmission of Nipah, Hendra, and perhaps SARS coronavirus and Ebola virus to humans may involve intermediate amplification hosts, such as bats, pigs, horses, civets, and primates. Understanding the natural reservoir or introductory host, the amplifying host, the epidemic center, and at-risk human populations are crucial in the control of emerging zoonosis. Particular species of bats have now been shown to promote coevolution between specific viruses and bat hosts, with the risk of cross-infection being able to generate new viruses that are capable of jumping the transmammalian species barrier more efficiently.

Ann. N.Y. Acad. Sci. 1102: xi–xii (2007). © 2007 New York Academy of Sciences.
doi: 10.1196/annals.1408.011

This volume aims at understanding the biology of some of the newly emerging viruses and their evolution. The H5N1 outbreaks of 1997 and 2004 are early warnings of the emergence of a pandemic. Good understanding of the basic biology of this virus and its evolutionary profiles is the key to combating huge uncontrollable influenza A outbreaks. The SARS-coronavirus has been a warning signal of the emergence of new uncontrollable viruses that show easy human–human transmission. With the potential to spread to large-scale proportions, there is a lot to learn from this "dress rehearsal" for a pandemic virus. Nipah virus showed traditional bat to pig to human transmission, until recently, whereas in Bangladesh direct bat to human, and human to human transmission was observed with higher fatality rates than SARS and H5N1. This phenomenon clearly needs to be studied, and the biology of the virus needs to be better understood.

Human metapneumoviruses have now been detected worldwide, causing severe respiratory tract illnesses primarily in the very young, the elderly, and immunocompromised individuals. Animal models and reverse genetic techniques have resulted in the first vaccine candidates to be developed; however, considerable genetic and antigenic diversity has been observed for human metapneumovirus, and the implications of this diversity for vaccine development and virus epidemiology require further study of the basic biology of the virus. Ross River virus-induced joint and muscle inflammation studies have been the focus of many laboratories worldwide until recently, with the development of a mouse model for study of the mechanisms behind this phenomenon, as described in this volume.

As emphasized earlier, proper surveillance and analysis and high throughput detection are described for West Nile virus in this book. Also, the use of PCR followed by electrospray ionization mass spectrometry, as well as newer approaches to vaccine development, using plants as bioreactors for vaccines, are some of the latest concepts briefly introduced in this volume. Although by no means exhaustive, our hope is that this volume will introduce or update the readership on the latest developments in the respective fields described above.

—Sunil K. Lal

International Centre for Genetic Engineering and Biotechnology
New Delhi, India

Biology of Influenza A Virus

TIMOTHY K. W. CHEUNG AND LEO L. M. POON

Department of Microbiology, Queen Mary Hospital, University of Hong Kong, Hong Kong, China

ABSTRACT: The outbreaks of avian influenza A virus in poultry and humans over the last decade posed a pandemic threat to human. Here, we discuss the basic classification and the structure of influenza A virus. The viral genome contains eight RNA viral segments and the functions of viral proteins encoded by this genome are described. In addition, the RNA transcription and replication of this virus are reviewed.

KEYWORDS: acidic polymerase protein (PA) avian influenza; basic polymerase protein 1 (PB1) basic polymerase protein 2 (PB2) genomic RNAs; vRNA; influenza A; *Orthomyxoviridae*; virology

INTRODUCTION

Influenza A viruses are medically important viral pathogens still causing significant mortality and morbidity throughout the world. By far the most catastrophic impact of influenza during the last century was the pandemic of Spanish flu in 1918, which cost more than 40 million lives.[1] The outbreaks of avian influenza H5N1/1997 and H5N1/2004 in the geographical region are reminders of the unpredictability of this pathogen and the potential for emergence of new pandemic viruses.[2,3] Thus, there remains a need for a better control of this infectious agent. A solid understanding of the underlying biology of influenza is therefore a key to better influenza pandemic preparedness. In this article, we will discuss some of the basic virology of the virus

ORTHOMYXOVIRUSES

Influenza A virus is classified within the family of *Orthomyxoviridae* (from the Greek *orthos*, meaning "standard, correct," and *myxa*, meaning "mucus"). Orthomyxoviruses are negative-sense, single-stranded, enveloped ribonucleic acid (RNA) viruses with a segmented genome. The genomic RNAs (vRNA)

Address for correspondence: Dr. Leo L. M. Poon, Department of Microbiology, University of Hong Kong, Queen Mary Hospital, Pokfulam, Hong Kong, China. Voice: 852-2819-9943; fax: 852-2855-1241.

llmpoon@hkucc.hku.hk

Ann. N.Y. Acad. Sci. 1102: 1–25 (2007).
doi: 10.1196/annals.1408.001

TABLE 1. Comparison among different classes of influenza viruses (Influenza A, B, and C viruses)[5]

Characteristics	Influenza A	Influenza B	Influenza C
Number of gene segment	8	8	7
Surface glycoprotein(s)	HA and NA	HA and NA	HEF (Hemagglutinin-Esterase-Fusion)
Host range	Wide (humans, pigs, horses, whales, seals, and birds)	Humans and seals	Mainly humans, but also found in swine

function as templates for messenger RNAs (mRNA) and complementary RNAs (cRNA) syntheses. There are four genera in the family of *Orthomyxoviridae*: Influenza virus A, Influenza virus B, Influenza virus C, and Thogotovirus. However, recent sequence analysis of infectious salmon anemia virus has suggested a new genus within the family of *Orthomyxoviridae*.[4]

CLASSIFICATION OF INFLUENZA VIRUSES

The influenza A, B, and C viruses can be distinguished on the basis of antigenic differences between their nucleoproteins (NP) and matrix proteins (M).[5] Both influenza A and B viruses contain 8 RNA genomic segments, whereas influenza C virus contains only 7 RNA genomic segments[6] (TABLE 1). All of these viruses can naturally infect humans. However, only influenza A virus has been responsible for all influenza pandemics.[7]

Based on the antigenic variation of the hemagglutinin (HA) and neuraminidase (NA) surface glycoproteins, influenza A viruses are further subdivided into different subtypes. Currently, there are 16 known subtypes of HA[8] and 9 known subtypes of NA.[9] Interestingly, each of these subtypes can be isolated from aquatic birds, suggesting avian species are the natural hosts of influenza viruses.[10] Of these viruses, only H1N1, H2N2, H3N2, H5N1, H7N7, and H9N2 subtypes have been isolated from human,[5,11–16] indicating that there is likely to be a host restriction for influenza viruses.

STRUCTURE OF THE INFLUENZA A VIRUS

Influenza virus is an enveloped virus and the virions are pleomorphic, with shapes ranging from small spherical to long filamentous. The viral morphology is a genetic trait and several viral proteins (HA, NA, M1, and M2) are known to have effects on the morphology of influenza virus particles.[17–22] Roberts and Compans[23] further demonstrated that the nature of the host cells also determines the morphology of influenza virus particles.

The influenza A viral particle contains a lipid envelope, which is derived from the host's cell membrane during the viral budding process. Three viral proteins, HA, NA, and M2, are embedded in the lipid envelope.[5] HA and NA are spike glycoproteins and they are anchored in the lipid bilayer by the short sequences of hydrophobic amino acids. Electron micrographs of influenza virus show that the HA spike and the NA spike are rod-shaped and mushroom-shaped, respectively. The HA is a homotrimer, which is responsible for the receptor binding and membrane fusion. The NA is a homotetramer whose function is to destroy receptors by hydrolyzing sialic acid groups from glycoproteins and to release the viral progeny. M2 protein is an integral membrane homotetramer, which functions as an ion channel for the acidification of the interior of the viral particle during viral infection.[24,25] Under the viral lipid envelope there is a M1 protein layer.[26]

Inside the virion, all eight vRNA segments are bound to the NP and to the influenza virus RNA polymerases to form ribonucleoprotein (RNP) complexes.[27] Apart from M1, NP is the most abundant protein in the virion and it is thought to associate with the phosphate-sugar backbone of the vRNA in a sequence-independent manner.[28] Each NP monomer interacts with approximately 20 nucleotides of the vRNA.[5] The RNA polymerase complex is composed of three polymerase subunits (PB2, PB1, and PA). Electron micrographs of isolated RNPs indicated that both ends of the vRNA interact with each other to form a circular or supercoil structure[29,30] and that the RNA polymerase interacts with one end of the RNP,[31] suggesting that the RNA polymerase interacts with both ends of the vRNA within the viral particle.[32] The NS2 protein is also present in virions in low amounts.[33,34] It appears to function as a nuclear export protein for vRNA in infected cells.[35]

GENOMIC ORGANIZATION OF THE INFLUENZA A VIRUS

The genome of the influenza A virus contains eight segments.[36] Viral mRNAs from segments 1 and 3 to 6 are monocistronic. Viral mRNAs from segment 2 of some viral isolates contain an alternative open reading frame. In contrast, viral mRNAs derived from segments 7 or 8 can undergo alternative splicing for protein expressions.[5] The sizes of the viral RNA segments and the proteins encoded are summarized in TABLE 2. Of these proteins, only the PB1-F2 protein from segment 2 (PB1 segment) and NS1 protein from segment 8 (NS segment) are nonstructural proteins.

Segment 1—Basic Polymerase Protein 2 (PB2)

Segment 1 of influenza A virus encodes one of the influenza viral polymerase subunits, PB2. It is widely accepted that PB2, PB1, PA, and NP form the

TABLE 2. The genome RNA and the corresponding encoded proteins of the influenza A/PR/8/34 virus

vRNA segment	vRNA length (bps)	Encoded protein(s)
1	2341	PB2
2	2341	PB1 & PB1-F2*
3	2233	PA
4	1778	HA
5	1565	NP
6	1413	NA
7	1027	M1 and M2**
8	890	NS1 and NS2(NEP)**

*Generated by an alternative open reading frame of the mRNA of PB1.
** Generated by splice mRNA.

minimum set of proteins required for viral transcription and replication.[37–39] PB2 contains a nuclear localization signal[40,41] and it is transported into the nucleus of infected cells for viral transcription and replication.[42] PB2 is an important protein for generating the cap structure for viral mRNAs. Studies of the influenza viral polymerase demonstrated that the PB2 subunit is a cap-binding protein.[43–46] Further analyses of this protein indicated that the cap-binding site is probably located near the carboxyl terminus.[47–49] Besides, PB2 has endonuclease activity[50,51] and it uses host mRNAs to generate cap primers for viral mRNA synthesis.[52–54] Consistent with this view, Nakagawa *et al.*[55] demonstrated that a cell line that expresses PB1, PA, and NP, but not PB2, can synthesize transcription products lacking 5′ cap structures, showing that PB2 is an important polymerase subunit for cap-snatching. Although the cap-snatching mechanism is not entirely understood, the 5′ and 3′ end of the vRNA template are shown to stimulate cap-snatching[56–60] and some of the aromatic residues in PB2 are shown to be critical for the process.[61]

Several studies indicated that PB2, PB1, and PA form a polymerase complex for viral transcription and replication. Immunoprecipitation assays on influenza viral polymerase demonstrated that PB2 is associated with the PB1 subunit.[62] Analysis of deletion mutants of PB2 indicated that the amino terminus of this protein is a binding site for PB1.[63] More recently, functional analysis of PB2 protein has shown that this polymerase subunit contains a novel binding site for PB1 subunit and two regions for binding nucleoprotein (NP) with regulatory interactions potential.[64]

Apart from the above biological functions, PB2 is also suggested to be a major determinant in controlling the pathogenicity of influenza A virus.[65] Using reverse genetics techniques, an introduction of a mutation at position 627 in the PB2 protein was shown to alter the virulence of H5N1/97 viruses in mice.[65]

Segment 2—Basic Polymerase Protein 1 (PB1)

The PB1 RNA polymerase subunit is encoded by segment 2. Several lines of evidence indicated that PB1, itself, is an RNA polymerase. Firstly, photochemical cross-linking assays demonstrated that both the substrate,[66] the elongated RNA product,[44] and the vRNA template[67–70] could be cross-linked to PB1, suggesting that PB1 carries the site for RNA polymerization. Secondly, amino acid sequence comparison with other RNA polymerases showed that the PB1 contains the four conserved motifs of RNA-dependent RNA polymerases[71] and mutations in these motifs abolished the polymerase activity.[72] Thirdly, nuclear extracts from cells expressing PB1 protein alone would transcribe model vRNA templates.[73]

Several studies have described the functional domains of PB1 involved in interaction with the other polymerase subunits. Immunoprecipitation studies of the influenza virus RNA polymerase suggested that PB1 contains independent binding sites for PB2 and PA.[62] Deletion mutant analyses of PB1 suggested that the amino- and carboxyl-termini of PB1 are binding sites for the PA and PB2 polymerase subunits, respectively.[63,74] The nuclear localization signal of PB1 was mapped to a region near the amino terminus.[75] The PB1 subunit plays a key role in both the assembly of three polymerase protein subunits and the catalytic function of RNA polymerization. Recently, Honda *et al.*[37] proposed that the catalytic specificity of PB1 subunit is modulated to the transcriptase by binding PB2 or the replicase by interaction with PA.

Segment 3—Acidic Polymerase Protein (PA)

The PA protein is encoded by segment 3 and is the smallest subunit of the influenza RNA polymerase complex. Like the other influenza viral polymerase subunits, it contains nuclear localization signals[76] required for transport into the nucleus.[42] PA is known to be essential for viral transcription and replication[38,39] and mutations near the carboxyl terminus inhibit transcription.[77] Fodor *et al.*[78,79] demonstrated that a single amino acid mutation in the PA subunit of the influenza virus RNA polymerase inhibits endonucleolytic cleavage of capped RNAs and promotes the generation of defective interfering RNAs. Furthermore, Fodor and Smith[80] illustrated that the PA subunit is required for efficient nuclear accumulation of the PB1 subunit of the influenza A virus RNA polymerase complex. Besides, amino acid sequence comparison with other known proteins suggested that the PA has helicase and ATP-binding activities.[47] However, the exact functions of PA are still poorly understood.

Interestingly, PA is found to induce proteolysis in infected cells,[81] but this property is not related to any known viral function and the significance of these findings is yet to be determined. Functional analysis of PA deletion mutants identified the amino-terminal one-third of this protein to be responsible for the protease activity.[81] When PA is expressed in cells in the absence of the other polymerase subunits, it induces a general proteolysis of both viral and cellular coexpressed proteins.[81] This might explain the failure to establish a PA-expressing cell line.[82] In addition, the proteolytic activity of PA seems to require the nuclear localization signals of the PA,[83] suggesting that nuclear transport is required for proteolytic activity. This might also explain the observations that the localization of PA in the nucleus is closely correlated with chromatin condensation and aberrant nuclear morphology.[84] Sanz-Ezquerro *et al.*[85] demonstrated that PA is a phosphorylated protein. Thus, the biological functions of PA protein might be regulated by a phosphorylation process.

Segment 4—Hemagglutinin (HA)

Segment 4 of influenza A virus encodes the HA. In viral particles, HA proteins associate as homotrimers. It is responsible for the binding of viral particles to sialic acid-containing receptors on the cell surface.[5,86] It also mediates the fusion of the viral and cellular membrane.[5,86] In addition, it is also the major target for neutralizing antibodies.[87] The HA is synthesized as a precursor polypeptide, HA_0.[5,86] This precursor polypeptide is post-translationally cleaved into two disulphide-linked subunits, HA_1 and HA_2. The cleavage of the HA_0 is a prerequisite for viral infectivity. This process liberates the "fusion peptide" at the amino terminus of HA_2 for membrane fusion. In addition, this cleavage also allows the native HA molecule to undergo a conformational change, a process that is triggered by an acidic environment and is essential for membrane fusion.[88] In general, the HA_0 is believed to be cleaved by trypsin-like proteases extracellularly. However, the presence of multiple basic amino acids within the cleavage site allows the protein to be cleaved by intracellular proteases, for example, furin,[89] which are ubiquitously expressed in most tissues. Hence, influenza viruses containing HA with multiple basic amino acids near the cleavage site are regarded as highly pathogenic and deletion of the multiple basic amino acids in the connective peptide could reduce the virulence of the highly pathogenic viruses.[90]

The HA receptor-binding site is a pocket that locates at the globular head domain of the HA_1. The amino acid sequence in the pocket region controls receptor-binding specificity.[91,92] For example, an HA molecule that contains a glutamine residue at position 226 prefers to bind to α2,3-linked sialic acid. Whereas, an HA molecule with a leucine residue at position 226 prefers to bind to α2,6-linked sialic acid.[5]

Segment 5—Nucleoprotein (NP)

NP is encoded by segment 5. The protein is a phosphorylated basic protein and has a net positive charge at neutral pH.[93,94] It is one of the essential components for transcription and replication.[38,39] The amino terminus of NP protein contains an RNA-binding domain. As a result, it has been suggested that the NP encapsidates the viral RNA in a sequence nonspecific manner.[95,96] Since the NP only binds to the backbone of the viral RNA, dissociation of the NP from the RNA template is not thought to be required for viral transcription and replication.[28] A temperature-sensitive mutant of NP has been shown to fail to synthesize complementary RNA (cRNA), but not mRNA, at the nonpermissive temperature.[97] Thus, this suggested that NP has a role in controlling the "switching" of RNA polymerase from transcription to replication. However, the mechanism by which NP might control such switching is not understood. Based on *in vitro* antibody depletion experiments, Shapiro and Krug[97] suggested that the NP binds to the RNA template and acts as an anti-termination factor for replication.[97,98] In contrast, based on immunoprecipitation assays and mutational analyses, Biswas *et al.*[99] and Mena *et al.*[100] suggested that the NP somehow regulates viral RNA replication by interacting with the viral polymerase subunits. However, others suggested that the transcription and replication of influenza are controlled via other mechanisms.[101] The NP has also been shown to be important for vRNA nuclear transport.[102–104] It contains nuclear localization signal(s)[105] and is a shuttle protein.[104,105] During the early stage of viral infection, the transport of incoming vRNPs from the viral particle into the nucleus is believed to be mediated by NP.[102–104] Whereas, in the late infection stage, progeny vRNAs associated with NP, M1, and NS2 are exported to the cytoplasm for viral packaging.[102,103,106]

Segment 6—Neuraminidase (NA)

Segment 6 of influenza A virus encodes the NA. The three-dimensional structure of the NA revealed that the NA monomer consists of four domains: a box-shaped globular head, a thin stalk, a transmembrane domain, and a cytoplasmic domain.[107] The NA is a surface glycoprotein and the glycosylation of the NA might be an important determinant (but not the sole determinant) of the neurovirulence of influenza viruses.[108] It exists as a homotetramer,[109] which has receptor-destroying activity to cleave the α–ketosidic linkage between a terminal sialic acid and an adjacent D-galactose or D-galactosamine residue.[110] The function of NA activity in the influenza virus life cycle is still not entirely clear. Liu *et al.*[111] demonstrated that a NA-deficient virus is infectious in cell culture and in mice, suggesting that the NA molecule is not required for viral entry, replication, and assembly. However, when the NA activity is inhibited, progeny viral particles attach to each other and/or to the cell surface to form large aggregates.[112,113] Therefore, it is generally assumed that NA has an important role in releasing progeny viral particles from infected cells. Recent studies have indicated that the conserved cytoplasmic tail of NA might control virion morphology and virulence.[20,21,114]

Segment 7—Matrix Proteins (M1 and M2)

Segment 7 of influenza A virus encodes two proteins, M1 and M2.[115] The M1 protein is a colinear transcription product of segment 7. In contrast, the M2 protein is encoded by the spliced mRNA of segment 7.

M1 Protein

In the viral particle, the M1 protein forms a layer to separate the RNP from the viral membrane[26,116] and it interacts with both the vRNA and protein components of RNP in assembly and disassembly of influenza A viruses.[117] The M1 protein is reported to have several functions for the virus. First, it binds to RNA in a sequence nonspecific manner[118–120] and inhibits viral transcription.[121] It contains a nuclear localization signal[122] and seems to regulate vRNP nuclear transport.[102] When it binds to vRNP, it promotes vRNP nuclear export and inhibits vRNP nuclear import.[102,123] Several studies demonstrated the nuclear export of viral ribonucleoproteins is associated with the matrix protein. It has been proposed that the vRNA and M1 protein together promote the self-assembly of influenza virus NP into the typical quaternary helical structure of the vRNP and the interaction of NP with vRNA and M1 in a system devoid of other viral proteins may lead to translocation of vRNP from the nucleus to the cytoplasm.[124] It was previously reported that M1 protein synthesized at high temperature (41°C) is unable to interact with vRNP and the transfer of vRNP into the cytoplasm from the nucleus is blocked, suggesting that the association between M1 and vRNP is essential for the nuclear export of vRNP.[125] Hirayama *et al.*[126] found that the induction of heat shock protein 70 (HSP70) by prostaglandin A1 (PGA1) at 37°C caused the suppression of virus production by preventing M1 protein from associating with vRNP and thus inhibiting the nuclear export of viral proteins. In addition, the M1 protein binds to the cell membrane[127] and seems to have an effect on viral assembly, budding, and viral morphology.[18,19,22,128–130] Structural analysis of a knockout mutant of influenza virus M1 protein indicated that the amino-terminal domain of this protein carries the nuclear localization sequence (NLS) motif and is important for membrane binding, self-polymerization, and nuclear export of vRNPs.[131] Recently, it has been suggested that the matrix protein has a potential to bind and inhibit the amidolytic activity of the viral RNA polymerase PA subunit.[132]

M2 Protein

The M2 protein is an integral membrane protein[133] and exists as a disulphide-bonded homotetramer.[134,135] The M2 tetramer has ion channel activity[24,25,136] for pH regulation. Takeuchi *et al.*[137] suggested that the amino acid residues in the transmembrane domain of the M2 protein, Histidine (His37) and Tryptophan (Trp41), are essential for the pH-regulated proton conductance. In the endosome of infected cells, the ion channel activity of M2 allows acidification of the interior of the incoming viral particle. The acidification of the viral particle is believed to be essential for viral replication, because it allows incoming vRNPs to dissociate from M1 proteins for nuclear import.[102,123] On the other hand, the ion channel activity of M2 is also reported to maintain a high pH in the Golgi vesicles so as to stabilize the native conformation of newly synthesized HA during the intracellular transport for viral assembly.[138] pH perturbations of acidic compartments via the influenza M2 proton channel affected the protein traffic along the secretory pathway.[139,140] It has been shown that M2 proteins, together with the M1 matrix proteins, are important determinants in filamentous viral particle formation.[22] However, recent study indicated that the M1 matrix protein would be the sole determinant to control the filamentous phenotype of influenza A virus.[18] Analyses of deletion mutants of M2 protein indicated that different domains of the ion channel take part in various viral processes. For example, the amino terminus of M2 protein is important for its incorporation into the virions,[141] whereas the carboxyl terminus is responsible for viral replication.[142] Moreover, several reports have shown that vaccine candidates for influenza viruses can be developed by targeting the extracellular domain,[143,144] as well as by deleting the transmembrane domain of M2 proteins.[145] The amino terminus of M2 protein has been shown to induce antibodies with inhibitory activity and protective immunity against the influenza virus replication.[146,147] A synthetic multiple antigenic vaccine that contains extracellular domain of M2 protein may induce influenza type A virus-specific resistance in mice.[148] Besides, Thomas *et al.*[149] reported that the phosphorylation status of a highly conserved phosphorylation site of the M2 protein (Serine 64) does not affect the intracellular transport of the protein in the infected cells. In addition, this study has also shown that the loss of phosphorylation on M2 proteins has not compromised the replication of influenza virus *in vivo*. Mutation analysis on M2 protein reported by Watanabe *et al.*[150] finds that influenza A virus without the M2 transmembrane domain, which is responsible for ion channel activity, can undergo multiple cycles of replication. In contrast, study of M2 mutant generated by Takeda *et al.*[151] suggests that M2 ion channel is essential for efficient viral replication in tissue culture.

Segment 8—Nonstructural Proteins (NS1 and NS2)

Segment 8 of influenza A virus encodes two proteins, NS1 and NS2.[152,153] The NS1 protein is a colinear transcription product of segment 8. In contrast, the NS2 is encoded by the spliced mRNA of segment 8.

NS1 Protein

The NS1 protein is the only nonstructural protein of influenza virus. It exists as an oligomer[154] and accumulates mainly in the nucleus.[155] The NS1 protein seems to regulate cellular and viral protein expression by binding to different RNA molecules. In many *in vitro* studies, NS1 has been shown to bind to a wide range of RNA molecules, such as poly(A)-containing cellular RNA,[156] vRNA,[157] vRNP,[158] double-stranded RNA (dsRNA),[159] and small nuclear RNA (snRNA).[160] It is also known to have inhibitory effects on splicing,[161,162] cellular mRNAs nuclear export,[156,161,163–165] cellular mRNA polyadenylation by interacting with the cellular 3′ end processing machinery[166,167] and dsRNA protein kinase (PKR) activation.[168] In addition, NS1 protein also appears to enhance viral protein expression by stimulating the translation of viral mRNA.[164,169,170] However, an influenza virus lacking the NS1 gene was generated in interferon-deficient cells,[171] suggesting that the NS1 protein is not absolutely essential for the viral life cycle in cell culture. Recently, NS1 protein of H5N1/97 was found to make the virus less susceptible to the antiviral effects of interferons and tumor necrosis factor alpha. In addition, the NS gene of H5N1/97 was shown to be a potent inducer of proinflammatory cytokines in human macrophages,[172] suggesting the unusual severity of human H5N1/97 disease might be due to the "cytokine storm" induced by the virus.

NS2 Protein

In early studies, it was believed that the NS2 protein was a nonstructural protein. However, further studies have indicated that NS2 is incorporated into viral particles in low amounts.[33,34] Based on studies of NS2 mutants, Odagiri *et al.*[173] suggested that NS2 plays a role in promoting normal replication of the genomic RNAs by an unknown mechanism. In addition, the carboxyl-terminal region of NS2 contains a M1 protein-binding site[34,174] suggesting that NS2 might regulate and cooperate with the function of M1. Based on the evidence that the NS2 protein contains a nuclear export signal and facilitates the vRNP export, O'Neill *et al.*[35] have proposed to rename this protein as NEP (viral nuclear export protein). Subsequence studies also confirmed that this protein is essential for vRNP nuclear export.[175]

TRANSCRIPTION AND REPLICATION OF INFLUENZA VIRUS RNA

The genome of influenza virus is negative-stranded RNA.[36] vRNAs are templates for both cRNA and mRNA syntheses. Unlike other negative-sense, single-stranded RNA viruses, the transcription and replication site for influenza virus genome is in the nucleus of infected cells.[176,177] All eight vRNA segments contain 12 and 13 conserved nucleotides at their 3′ and 5′ ends, respectively.[6,178,179] In addition, each RNA segment also contains 2 to 3 segment-specific nucleotides near each end. These RNA sequences are partially complementary and can form a panhandle structure.[6,178,179] Viral mRNA is a capped and polyadenylated transcript.[53,54,180] It is an incomplete copy of the vRNA, lacking approximately 17 bases of the complementary viral 3′ sequence.[181] In contrast, cRNA is a faithful copy of vRNA and acts as a template for further synthesis of vRNA.[182,183] Unlike vRNA and mRNAs, cRNAs are not transported to the cytoplasm during viral infection.[184] In general, it is believed that all the polymerase subunits (PB2, PB1, and PA) and NP are required for both viral transcription and replication.[38,39]

VIRAL TRANSCRIPTION OF INFLUENZA VIRUS

Transcription of mRNA is initiated by a capped RNA fragment, which is cleaved from host mRNA by a cap-snatching mechanism.[44,53,54] The PB2 polymerase subunit binds to the 5′ end of host cell mRNA and cleaves it at about 10–15 nucleotides downstream from the cap structure after predominantly an A or a G residue.[50] This specific cleavage requires the presence of a methylated cap structure in the RNA substrate[52] and the endonuclease activity of PB2 is stimulated by the vRNA.[56,57,70] Several capped RNA or dinucleotides (e.g., ApG) are known to be used as primers for transcription *in vitro*.[52,185–187] However, of these primers, the cap 1 structure (m^7GpppX^m) from mammalian cellular mRNA has been shown to be the preferred primer for viral transcription.[52] Interestingly, although the viral mRNAs also contain the cap 1 structure, they seem to be protected from endonucleolytic cleavage.[188] These observations suggested that the viral polymerase complex selectively uses host mRNA for viral mRNA synthesis.[188]

After endonuclease cleavage, the short-capped oligonucleotide is used by the viral polymerase as a primer for transcription. Initiation of viral transcription requires the interaction between the 5′ and 3′ conserved sequences of vRNA[68,189,190] and several different secondary structure models of these promoter sequences have been proposed.[28,68,189,191] Early studies suggested that base pairing of the 3′ terminal nucleotide of the cleaved primer with the 3′ terminal U residue (i.e., position 1) of the vRNA template was not required for transcription initiation.[192] However, subsequently, it became clear that this interaction has an effect on the position of transcription initiation.[193]

Transcription is terminated at a track of 5–7 U residues approximately 17 nucleotides from the 5′ end of vRNA and a poly(A) tail is then added to the mRNA transcript.[181,194] RNPs isolated from viral particles can synthesize polyadenylated transcripts *in vitro*, indicating that the polyadenylation of influenza virus mRNA is a host-independent process.[195] Currently, it is generally believed that the viral RNA polymerase reiteratively copies the U-track at the 5′ end of vRNA to polyadenylate mRNAs[196] and several polyadenylation models have been proposed.[29,68,181,190,194,197] *In vivo* studies have indicated that the U-track is required for polyadenylation of viral mRNA.[190,198] Mutational analysis showed that polyadenylated mRNA synthesized by a recombinant influenza virus with mutated U-track to A-track at the 5′ end of vRNA was defective in nuclear export.[199] This result confirms that the poly(A) tail generated by U-track, together with the cap structure, are normally essential features for mRNA nuclear export[200,201] and mRNA stabilization.[202,203]

VIRAL REPLICATION OF INFLUENZA VIRUS

cRNA is a full-length copy of the vRNA and can be used as a template for vRNA synthesis. Unlike the products of transcription, which are capped at their 5′ end, the 5′ terminus of cRNA is pppA, suggesting that the synthesis of cRNA is a primer-independent process.[183] In addition, cRNA is not polyadenylated. Although both cellular and viral factors have been suggested to have crucial roles in the transcription–replication regulation,[99,184,204,205] the mechanism for the switching from transcription to replication is still poorly understood. Recently, Vreede *et al.*[101] demonstrated that there may be no switch regulating the initiation of RNA synthesis and proposed a model suggesting that nascent cRNA is degraded by host cell nucleases unless it is stabilized by newly synthesized viral RNA polymerase and NP proteins.

EXPRESSION OF INFLUENZA VIRAL PROTEINS

It is widely accepted that the replication and transcription of the influenza viral genome is a selective process.[182,206] In infected cells, the synthesis of each mRNA molecule is known to vary over the time course of infection. However, the factors that control viral gene expression are not entirely understood. In general, the amount of viral proteins synthesized in infected cells is largely dependent on the amount of the corresponding mRNA in infected cells.[182,184] Immediately after infection, primary transcription occurs.[182] In this stage, all eight mRNAs are synthesized in equimolar amounts. This is followed by the second transcription stage. The second transcription stage can be further subdivided into early and late phases. In the early phase of the secondary transcription, NS1 and NP vRNA are preferentially synthesized. As a consequence, NS1 and NP are the predominant viral proteins in infected cells at this stage.[182,184,206] The reason, however, for the preferential early expression of the NS1 and NP proteins is not known. It is possible that NP is required for the replication and transcription of viral RNA. NS1 might be required for the regulation of cellular gene expression. During the late phase, vRNAs are synthesized in equivalent amounts, as required for progeny virus genome. At this stage, the NS1 protein is synthesized in a reduced level.[182,184,206] In contrast, HA, NA, and M1 mRNAs are preferentially expressed.[182,184,206] In general, most of the capped and polyadenylated viral mRNAs would be transported from the nucleus to the cytoplasm for protein synthesis. On the other hand, the membrane-bounded proteins, such as HA, NA, and M2, would pass the secretory pathway at the *trans*-Golgi network for protein maturation. HA and NA proteins are post-translationally modified and transported to the cell surface for integration into the cell membrane.

NUCLEAR EXPORT OF VIRAL RIBONUCLEOPROTEINS (VRNPS)

Recent evidence suggests that the M1 protein is associated with the migration of ribonucleoproteins out of the nucleus for assembly into progeny viral particles in the cytoplasm.[102,123–125] However, it was found that NS2 protein might also be involved in the translocation of vRNP for viral packaging in the cytoplasm.[102,103] It is because M1 protein contains a nuclear localization domain[122] that allows it to enter to and exit from the nucleus to regulate vRNP nuclear transport.[102] When it binds to vRNP, it promotes vRNP nuclear export. Since the carboxyl-terminal region of NS2 contains an M1 protein-binding site,[34,174] it is suggested that NS2 might regulate and cooperate with the function of M1 by interacting with the M1 protein for assisting the nuclear export of vRNPs.

ACKNOWLEDGMENTS

We acknowledge funding from the National Institute of Allergy and Infectious Disease, USA (AI95357), the Research Grant Council of Hong Kong (HKU 7343/04M), and the University of Hong Kong (seed funding for new staff, 2004–2005).

REFERENCES

1. PALESE, P. 2004. Influenza: old and new threats. Nat. Med. **10:** S82–S87.
2. LI, K.S., Y. GUAN, J. WANG, *et al.* 2004. Genesis of a highly pathogenic and potentially pandemic H5N1 influenza virus in eastern Asia. Nature **430:** 209–213.
3. WEBBY, R.J. & R.G. WEBSTER. 2003. Are we ready for pandemic influenza? Science **302:** 1519–1522.
4. KROSSOY, B., I. HORDVIK, F. NILSEN, *et al.* 1999. The putative polymerase sequence of infectious salmon anemia virus suggests a new genus within the *Orthomyxoviridae*. J. Virol. **73:** 2136–2142.

5. LAMB, R.A. & R.M. KRUG. 2001. *Orthomyxoviridae*: the viruses and their replication. *In* Fields Virology. vol. 1. Fourth edition D.M. Knipe & P.M. Howley, Eds.: 1487–1531. Lippincott Williams and Wilkins. Philadelphia.
6. DESSELBERGER, U., V.R. RACANIELLO, J.J. ZAZRA & P. PALESE. 1980. The 3' and 5'-terminal sequences of influenza A, B and C virus RNA segments are highly conserved and show partial inverted complementarity. Gene **8:** 315–328.
7. POTTER, C.W. 1998. Chronicle of influenza pandemics. *In* Textbook of Influenza. K.G. Nicholson, R.G. Webster & A.J. Hay, Eds.: 3–18. Blackwell Science. Oxford, UK.
8. FOUCHIER, R.A., V. MUNSTER, A. WALLENSTEN, *et al.* 2005. Characterization of a novel influenza A virus hemagglutinin subtype (H16) obtained from black-headed gulls. J. Virol. **79:** 2814–2822.
9. LAVER, W.G., P.M. COLMAN, R.G. WEBSTER, *et al.* 1984. Influenza virus neuraminidase with hemagglutinin activity. Virology **137:** 314–323.
10. ITO, T. & Y. KAWAOKA. 1998. Avian influenza. *In* Textbook of Influenza. K.G. Nicholson, R.G. Webster & A.J. Hay, Eds.: 126–136. Blackwell Science. Oxford, UK.
11. CLAAS, E.C., A.D. OSTERHAUS, R. VAN BEEK, *et al.* 1998. Human influenza A H5N1 virus related to a highly pathogenic avian influenza virus. Lancet **351:** 472–477.
12. FOUCHIER, R.A., P.M. SCHNEEBERGER, F.W. ROZENDAAL, *et al.* 2004. Avian influenza A virus (H7N7) associated with human conjunctivitis and a fatal case of acute respiratory distress syndrome. Proc. Natl. Acad. Sci. USA **101:** 1356–1361.
13. GUAN, Y., K.F. SHORTRIDGE, S. KRAUSS & R.G. WEBSTER. 1999. Molecular characterization of H9N2 influenza viruses: were they the donors of the "internal" genes of H5N1 viruses in Hong Kong? Proc. Natl. Acad. Sci. USA **96:** 9363–9367.
14. GUAN, Y., J.S. PEIRIS, L.L. POON, *et al.* 2003. Reassortants of H5N1 influenza viruses recently isolated from aquatic poultry in Hong Kong SAR. Avian Dis. **47:** 911–913.
15. SUBBARAO, K., A. KLIMOV, J. KATZ, *et al.* 1998. Characterization of an avian influenza A (H5N1) virus isolated from a child with a fatal respiratory illness. Science **279:** 393–396.
16. YUEN, K.Y., P.K. CHAN, M. PEIRIS, *et al.* 1998. Clinical features and rapid viral diagnosis of human disease associated with avian influenza A H5N1 virus. Lancet **351:** 467–471.
17. BURLEIGH, L.M., L.J. CALDER, J.J. SKEHEL & D.A. STEINHAUER. 2005. Influenza A viruses with mutations in the m1 helix six domain display a wide variety of morphological phenotypes. J. Virol. **79:** 1262–1270.
18. ELLEMAN, C.J. & W.S. BARCLAY. 2004. The M1 matrix protein controls the filamentous phenotype of influenza A virus. Virology **321:** 144–153.
19. ENAMI, M. & K. ENAMI. 1996. Influenza virus hemagglutinin and neuraminidase glycoproteins stimulate the membrane association of the matrix protein. J. Virol. **70:** 6653–6657.
20. JIN, H., G.P. LESER, J. ZHANG & R.A. LAMB. 1997. Influenza virus hemagglutinin and neuraminidase cytoplasmic tails control particle shape. EMBO J. **16:** 1236–1247.
21. MITNAUL, L.J., M.R. CASTRUCCI, K.G. MURTI & Y. KAWAOKA. 1996. The cytoplasmic tail of influenza A virus neuraminidase (NA) affects NA incorporation

into virions, virion morphology, and virulence in mice but is not essential for virus replication. J. Virol. **70:** 873–879.
22. Roberts, P.C., R.A. Lamb & R.W. Compans. 1998. The M1 and M2 proteins of influenza A virus are important determinants in filamentous particle formation. Virology **240:** 127–137.
23. Roberts, P.C. & R.W. Compans. 1998. Host cell dependence of viral morphology. Proc. Natl. Acad. Sci. USA **95:** 5746–5751.
24. Pinto, L.H., L.J. Holsinger & R.A. Lamb. 1992. Influenza virus M2 protein has ion channel activity. Cell **69:** 517–528.
25. Wang, C., R.A. Lamb & L.H. Pinto. 1994. Direct measurement of the influenza A virus M2 protein ion channel activity in mammalian cells. Virology **205:** 133–140.
26. Ruigrok, R.W., L.J. Calder & S.A. Wharton. 1989. Electron microscopy of the influenza virus submembranal structure. Virology **173:** 311–316.
27. Lamb, R.A. & P.W. Choppin. 1983. The gene structure and replication of influenza virus. Annu. Rev. Biochem. **52:** 467–506.
28. Baudin, F., C. Bach, S. Cusack & R.W. Ruigrok. 1994. Structure of influenza virus RNP. Part I. Influenza virus nucleoprotein melts secondary structure in panhandle RNA and exposes the bases to the solvent. EMBO J. **13:** 3158–3165.
29. Hsu, M.T., J.D. Parvin, S. Gupta, *et al.* 1987. Genomic RNAs of influenza viruses are held in a circular conformation in virions and in infected cells by a terminal panhandle. Proc. Natl. Acad. Sci. USA **84:** 8140–8144.
30. Jennings, P.A., J.T. Finch, G. Winter & J.S. Robertson. 1983. Does the higher order structure of the influenza virus ribonucleoprotein guide sequence rearrangements in influenza viral RNA? Cell **34:** 619–627.
31. Murti, K.G., R.G. Webster & I.M. Jones. 1988. Localization of RNA polymerases on influenza viral ribonucleoproteins by immunogold labeling. Virology **164:** 562–566.
32. Klumpp, K., R.W. Ruigrok & F. Baudin. 1997. Roles of the influenza virus polymerase and nucleoprotein in forming a functional RNP structure. EMBO J. **16:** 1248–1257.
33. Richardson, J.C. & R.K. Akkina. 1991. NS2 protein of influenza virus is found in purified virus and phosphorylated in infected cells. Arch. Virol. **116:** 69–80.
34. Yasuda, J., S. Nakada, A. Kato, *et al.* 1993. Molecular assembly of influenza virus: association of the NS2 protein with virion matrix. Virology **196:** 249–255.
35. O'Neill, R.E., J. Talon & P. Palese. 1998. The influenza virus NEP (NS2 protein) mediates the nuclear export of viral ribonucleoproteins. EMBO J. **17:** 288–296.
36. Palese, P. 1977. The genes of influenza virus. Cell **10:** 1–10.
37. Honda, A., K. Mizumoto & A. Ishihama. 2002. Minimum molecular architectures for transcription and replication of the influenza virus. Proc. Natl. Acad. Sci. USA **99:** 13166–13171.
38. Huang, T.S., P. Palese & M. Krystal. 1990. Determination of influenza virus proteins required for genome replication. J. Virol. **64:** 5669–5673.
39. Perales, B. & J. Ortin. 1997. The influenza A virus PB2 polymerase subunit is required for the replication of viral RNA. J. Virol. **71:** 1381–1385.

40. MUKAIGAWA, J. & D.P. NAYAK. 1991. Two signals mediate nuclear localization of influenza virus (A/WSN/33) polymerase basic protein 2. J. Virol. **65:** 245–253.
41. PERALES, B., S. DE LA LUNA, I. PALACIOS & J. ORTIN. 1996. Mutational analysis identifies functional domains in the influenza A virus PB2 polymerase subunit. J. Virol. **70:** 1678–1686.
42. JONES, I.M., P.A. REAY & K.L. PHILPOTT. 1986. Nuclear location of all three influenza polymerase proteins and a nuclear signal in polymerase PB2. EMBO J. **5:** 2371–2376.
43. BLAAS, D., E. PATZELT & E. KUECHLER. 1982. Identification of the cap binding protein of influenza virus. Nucleic Acids Res. **10:** 4803–4812.
44. BRAAM, J., I. ULMANEN & R.M. KRUG. 1983. Molecular model of a eucaryotic transcription complex: functions and movements of influenza P proteins during capped RNA-primed transcription. Cell **34:** 609–618.
45. ULMANEN, I., B.A. BRONI & R.M. KRUG. 1981. Role of two of the influenza virus core P proteins in recognizing cap 1 structures (m7GpppNm) on RNAs and in initiating viral RNA transcription. Proc. Natl. Acad. Sci. USA **78:** 7355–7359.
46. ULMANEN, I., B. BRONI & R.M. KRUG. 1983. Influenza virus temperature-sensitive cap (m7GpppNm)-dependent endonuclease. J. Virol. **45:** 27–35.
47. DE LA LUNA, S., C. MARTINEZ & J. ORTIN. 1989. Molecular cloning and sequencing of influenza virus A/Victoria/3/75 polymerase genes: sequence evolution and prediction of possible functional domains. Virus Res. **13:** 143–155.
48. MASUNAGA, K., K. MIZUMOTO, H. KATO, *et al.* 1999. Molecular mapping of influenza virus RNA polymerase by site-specific antibodies. Virology **256:** 130–141.
49. YAMANAKA, K., A. ISHIHAMA & K. NAGATA. 1990. Reconstitution of influenza virus RNA-nucleoprotein complexes structurally resembling native viral ribonucleoprotein cores. J. Biol. Chem. **265:** 11151–11155.
50. PLOTCH, S.J., M. BOULOY, I. ULMANEN & R.M. KRUG. 1981. A unique cap(m7GpppXm)-dependent influenza virion endonuclease cleaves capped RNAs to generate the primers that initiate viral RNA transcription. Cell **23:** 847–858.
51. SHI, L., D.F. SUMMERS, Q. PENG & J.M. GALARZ. 1995. Influenza A virus RNA polymerase subunit PB2 is the endonuclease which cleaves host cell mRNA and functions only as the trimeric enzyme. Virology **208:** 38–47.
52. BOULOY, M., S.J. PLOTCH & R.M. KRUG. 1980. Both the 7-methyl and the 2'-O-methyl groups in the cap of mRNA strongly influence its ability to act as primer for influenza virus RNA transcription. Proc. Natl. Acad. Sci. USA **77:** 3952–3956.
53. KRUG, R.M., B.A. BRONI & M. BOULOY. 1979. Are the 5′ ends of influenza viral mRNAs synthesized *in vivo* donated by host mRNAs? Cell **18:** 329–334.
54. PLOTCH, S.J., M. BOULOY & R.M. KRUG. 1979. Transfer of 5′-terminal cap of globin mRNA to influenza viral complementary RNA during transcription *in vitro*. Proc. Natl. Acad. Sci. USA **76:** 1618–1622.
55. NAKAGAWA, Y., N. KIMURA, T. TOYODA, *et al.* 1995. The RNA polymerase PB2 subunit is not required for replication of the influenza virus genome but is involved in capped mRNA synthesis. J. Virol. **69:** 728–733.
56. CIANCI, C., L. TILEY & M. KRYSTAL. 1995. Differential activation of the influenza virus polymerase via template RNA binding. J. Virol. **69:** 3995–3999.

57. HAGEN, M., T.D. CHUNG, J.A. BUTCHER & M. KRYSTAL. 1994. Recombinant influenza virus polymerase: requirement of both 5′ and 3′ viral ends for endonuclease activity. J. Virol. **68:** 1509–1515.
58. LEAHY, M.B., H.C. DOBBYN & G.G. BROWNLEE. 2001a. Hairpin loop structure in the 3′ arm of the influenza A virus virion RNA promoter is required for endonuclease activity. J. Virol. **75:** 7042–7049.
59. LEAHY, M.B., D.C. PRITLOVE, L.L. POON & G.G. BROWNLEE. 2001b. Mutagenic analysis of the 5′ arm of the influenza A virus virion RNA promoter defines the sequence requirements for endonuclease activity. J. Virol. **75:** 134–142.
60. LI, M.L., P. RAO & R.M. KRUG. 2001. The active sites of the influenza cap-dependent endonuclease are on different polymerase subunits. EMBO J. **20:** 2078–2086.
61. FECHTER, P., L. MINGAY, J. SHARPS, *et al.* 2003. Two aromatic residues in the PB2 subunit of influenza A RNA polymerase are crucial for cap binding. J. Biol. Chem. **278:** 20381–20388.
62. DIGARD, P., V.C. BLOK & S.C. INGLIS. 1989. Complex formation between influenza virus polymerase proteins expressed in *Xenopus* oocytes. Virology **171:** 162–169.
63. TOYODA, T., D.M. ADYSHEV, M. KOBAYASHI, *et al.* 1996. Molecular assembly of the influenza virus RNA polymerase: determination of the subunit-subunit contact sites. J. Gen. Virol. **77:** 2149–2157.
64. POOLE, E., D. ELTON, L. MEDCALF & P. DIGARD. 2004. Functional domains of the influenza A virus PB2 protein: identification of NP- and PB1-binding sites. Virology **321:** 120–133.
65. HATTA, M., P. GAO, P. HALFMANN & Y. KAWAOKA. 2001. Molecular basis for high virulence of Hong Kong H5N1 influenza A viruses. Science **293:** 1840–1842.
66. ASANO, Y. & A. ISHIHAMA. 1997. Identification of two nucleotide-binding domains on the PB1 subunit of influenza virus RNA polymerase. J. Biochem. **122:** 627–634.
67. FODOR, E., B.L. SEONG & G.G. BROWNLEE. 1993. Photochemical cross-linking of influenza A polymerase to its virion RNA promoter defines a polymerase binding site at residues 9 to 12 of the promoter. J. Gen. Virol. **74:** 1327–1333.
68. FODOR, E., D.C. PRITLOVE & G.G. BROWNLEE. 1994. The influenza virus panhandle is involved in the initiation of transcription. J. Virol. **68:** 4092–4096.
69. GONZALEZ, S. & J. ORTIN. 1999. Characterization of influenza virus PB1 protein binding to viral RNA: two separate regions of the protein contribute to the interaction domain. J. Virol. **73:** 631–637.
70. LI, M.L., B.C. RAMIREZ & R.M. KRUG. 1998. RNA-dependent activation of primer RNA production by influenza virus polymerase: different regions of the same protein subunit constitute the two required RNA-binding sites. EMBO J. **17:** 5844–5852.
71. POCH, O., I. SAUVAGET, M. DELARUE & N. TORDO. 1989. Identification of four conserved motifs among the RNA-dependent polymerase encoding elements. EMBO J. **8:** 3867–3874.
72. BISWAS, S.K. & D.P. NAYAK. 1994. Mutational analysis of the conserved motifs of influenza A virus polymerase basic protein 1. J. Virol. **68:** 1819–1826.
73. KOBAYASHI, M., T. TOYODA & A. ISHIHAMA. 1996. Influenza virus PB1 protein is the minimal and essential subunit of RNA polymerase. Arch. Virol. **141:** 525–539.

74. Gonzalez, S., T. Zurcher & J. Ortin. 1996. Identification of two separate domains in the influenza virus PB1 protein involved in the interaction with the PB2 and PA subunits: a model for the viral RNA polymerase structure. Nucleic Acids Res. **24:** 4456–4463.
75. Nath, S.T. & D.P. Nayak. 1990. Function of two discrete regions is required for nuclear localization of polymerase basic protein 1 of A/WSN/33 influenza virus (H1 N1). Mol. Cell. Biol. **10:** 4139–4145.
76. Nieto, A., S. de la Luna, J. Barcena, *et al.* 1994. Complex structure of the nuclear translocation signal of influenza virus polymerase PA subunit. J. Gen. Virol. **75:** 29–36.
77. Zurcher, T., S. de la Luna, J.J. Sanz-Ezquerro, *et al.* 1996. Mutational analysis of the influenza virus A/Victoria/3/75 PA protein: studies of interaction with PB1 protein and identification of a dominant negative mutant. J. Gen. Virol. **77:** 1745–1749.
78. Fodor, E., M. Crow, L.J. Mingay, *et al.* 2002. A single amino acid mutation in the PA subunit of the influenza virus RNA polymerase inhibits endonucleolytic cleavage of capped RNAs. J. Virol. **76:** 8989–9001.
79. Fodor, E., L.J. Mingay, M. Crow, *et al.* 2003. A single amino acid mutation in the PA subunit of the influenza virus RNA polymerase promotes the generation of defective interfering RNAs. J. Virol. **77:** 5017–5020.
80. Fodor, E. & M. Smith. 2004. The PA subunit is required for efficient nuclear accumulation of the PB1 subunit of the influenza A virus RNA polymerase complex. J. Virol. **78:** 9144–9153.
81. Sanz-Ezquerro, J.J., S. de la Luna, J. Ortin & A. Nieto. 1995. Individual expression of influenza virus PA protein induces degradation of coexpressed proteins. J. Virol. **69:** 2420–2426.
82. Stranden, A.M., P. Staeheli & J. Pavlovic. 1993. Function of the mouse Mx1 protein is inhibited by overexpression of the PB2 protein of influenza virus. Virology **197:** 642–651.
83. Sanz-Ezquerro, J.J., T. Zurcher, S. de la Luna, *et al.* 1996. The amino-terminal one-third of the influenza virus PA protein is responsible for the induction of proteolysis. J. Virol. **70:** 1905–1911.
84. Nieto, A., S. de la Luna, J. Barcena, *et al.* 1992. Nuclear transport of influenza virus polymerase PA protein. Virus Res. **24:** 65–75.
85. Sanz-Ezquerro, J.J., J. Fernandez Santaren, T. Sierra, *et al.* 1998. The PA influenza virus polymerase subunit is a phosphorylated protein. J. Gen. Virol. **79:** 471–478.
86. Steinhauer, D.A. & S.A. Wharton. 1998. Structure and function of the haemagglutinin. *In* Textbook of Influenza. K.G. Nicholson, R.G. Webster & A.J. Hay, Eds.: 54–64. Blackwell Science. Oxford, UK.
87. Staudt, L.M. & W. Gerhard. 1983. Generation of antibody diversity in the immune response of BALB/c mice to influenza virus hemagglutinin. I. Significant variation in repertoire expression between individual mice. J. Exp. Med. **157:** 687–704.
88. Skehel, J.J., P.M. Bayley, E.B. Brown, *et al.* 1982. Changes in the conformation of influenza virus hemagglutinin at the pH optimum of virus-mediated membrane fusion. Proc. Natl. Acad. Sci. USA **79:** 968–972.
89. Horimoto, T., K. Nakayama, S.P. Smeekens & Y. Kawaoka. 1994. Proprotein-processing endoproteases PC6 and furin both activate hemagglutinin of virulent avian influenza viruses. J. Virol. **68:** 6074–6078.

90. WEBBY, R.J., D.R. PEREZ, J.S. COLEMAN, *et al.* 2004. Responsiveness to a pandemic alert: use of reverse genetics for rapid development of influenza vaccines. Lancet **363:** 1099–1103.
91. CONNOR, R.J., Y. KAWAOKA, R.G. WEBSTER & J.C. PAULSON. 1994. Receptor specificity in human, avian, and equine H2 and H3 influenza virus isolates. Virology **205:** 17–23.
92. VINES, A., K. WELLS, M. MATROSOVICH, *et al.* 1998. The role of influenza A virus hemagglutinin residues 226 and 228 in receptor specificity and host range restriction. J. Virol. **72:** 7626–7631.
93. KISTNER, O., K. MULLER & C. SCHOLTISSEK. 1989. Differential phosphorylation of the nucleoprotein of influenza A viruses. J. Gen. Virol. **70:** 2421–2431.
94. WINTER, G. & S. FIELDS. 1981. The structure of the gene encoding the nucleoprotein of human influenza virus A/PR/8/34. Virology **114:** 423–428.
95. ALBO, C., A. VALENCIA & A. PORTELA. 1995. Identification of an RNA binding region within the N-terminal third of the influenza A virus nucleoprotein. J. Virol. **69:** 3799–3806.
96. KOBAYASHI, M., T. TOYODA, D.M. ADYSHEV, *et al.* 1994. Molecular dissection of influenza virus nucleoprotein: deletion mapping of the RNA binding domain. J. Virol. **68:** 8433–8436.
97. SHAPIRO, G.I. & R.M. KRUG. 1988. Influenza virus RNA replication *in vitro*: synthesis of viral template RNAs and virion RNAs in the absence of an added primer. J. Virol. **62:** 2285–2290.
98. BEATON, A.R. & R.M. KRUG. 1986. Transcription antitermination during influenza viral template RNA synthesis requires the nucleocapsid protein and the absence of a 5′ capped end. Proc. Natl. Acad. Sci. USA **83:** 6282–6286.
99. BISWAS, S.K., P.L. BOUTZ & D.P. NAYAK. 1998. Influenza virus nucleoprotein interacts with influenza virus polymerase proteins. J. Virol. **72:** 5493–5501.
100. MENA, I., E. JAMBRINA, C. ALBO, *et al.* 1999. Mutational analysis of influenza A virus nucleoprotein: identification of mutations that affect RNA replication. J. Virol. **73:** 1186–1194.
101. VREEDE, F.T., T.E. JUNG & G.G. BROWNLEE. 2004. Model suggesting that replication of influenza virus is regulated by stabilization of replicative intermediates. J. Virol. **78:** 9568–9572.
102. MARTIN, K. & A. HELENIUS. 1991. Nuclear transport of influenza virus ribonucleoproteins: the viral matrix protein (M1) promotes export and inhibits import. Cell **67:** 117–130.
103. O'NEILL, R.E., R. JASKUNAS, G. BLOBEL, *et al.* 1995. Nuclear import of influenza virus RNA can be mediated by viral nucleoprotein and transport factors required for protein import. J. Biol. Chem. **270:** 22701–22704.
104. WHITTAKER, G., M. BUI & A. HELENIUS. 1996. Nuclear trafficking of influenza virus ribonucleoproteins in heterokaryons. J. Virol. **70:** 2743–2756.
105. NEUMANN, G., M.R. CASTRUCCI & Y. KAWAOKA. 1997. Nuclear import and export of influenza virus nucleoprotein. J. Virol. **71:** 9690–9700.
106. ELTON, D., M. SIMPSON-HOLLEY, K. ARCHER, *et al.* 2001. Interaction of the influenza virus nucleoprotein with the cellular CRM1-mediated nuclear export pathway. J. Virol. **75:** 408–419.
107. VARGHESE, J.N. & P.M. COLMAN. 1991. Three-dimensional structure of the neuraminidase of influenza virus A/Tokyo/3/67 at 2.2 Å resolution. J. Mol. Biol. **221:** 473–486.

108. LI, S., J. SCHULMAN, S. ITAMURA & P. PALESE. 1993. Glycosylation of neuraminidase determines the neurovirulence of influenza A/WSN/33 virus. J. Virol. **67:** 6667–6673.
109. HAUSMANN, J., E. KRETZSCHMAR, W. GARTEN & H.D. KLENK. 1997. Biosynthesis, intracellular transport and enzymatic activity of an avian influenza A virus neuraminidase: role of unpaired cysteines and individual oligosaccharides. J. Gen. Virol. **78:** 3233–3245.
110. COLMAN, P.M. 1998. Structure and function of the haemagglutinin. *In* Textbook of Influenza. K.G. Nicholson, R.G. Webster & A.J. Hay, Eds.: 65–73. Blackwell Science. Oxford, UK.
111. LIU, C., M.C. EICHELBERGER, R.W. COMPANS & G.M. AIR. 1995. Influenza type A virus neuraminidase does not play a role in viral entry, replication, assembly, or budding. J. Virol. **69:** 1099–1106.
112. PALESE, P., K. TOBITA, M. UEDA & R.W. COMPANS. 1974. Characterization of temperature sensitive influenza virus mutants defective in neuraminidase. Virology **61:** 397–410.
113. PALESE, P. & R.W. COMPANS. 1976. Inhibition of influenza virus replication in tissue culture by 2-deoxy-2,3-dehydro-N-trifluoroacetylneuraminic acid (FANA): mechanism of action. J. Gen. Virol. **33:** 159–163.
114. BILSEL, P., M.R. CASTRUCCI & Y. KAWAOKA. 1993. Mutations in the cytoplasmic tail of influenza A virus neuraminidase affect incorporation into virions. J. Virol. **67:** 6762–6767.
115. SHIH, S.R., M.E. NEMEROFF & R.M. KRUG. 1995. The choice of alternative 5′ splice sites in influenza virus M1 mRNA is regulated by the viral polymerase complex. Proc. Natl. Acad. Sci. USA **92:** 6324–6328.
116. RUIGROK, R.W., A. BARGE, P. DURRER, *et al.* 2000. Membrane interaction of influenza virus M1 protein. Virology **267:** 289–298.
117. YE, Z., T. LIU, D.P. OFFRINGA, *et al.* 1999. Association of influenza virus matrix protein with ribonucleoproteins. J. Virol. **73:** 7467–7473.
118. ELSTER, C., E. FOUREST, F. BAUDIN, *et al.* 1994. A small percentage of influenza virus M1 protein contains zinc but zinc does not influence *in vitro* M1-RNA interaction. J. Gen. Virol. **75:** 37–42.
119. WAKEFIELD, L. & G.G. BROWNLEE. 1989. RNA-binding properties of influenza A virus matrix protein M1. Nucleic Acids Res. **17:** 8569–8580.
120. YE, Z.P., N.W. BAYLOR & R.R. WAGNER. 1989. Transcription-inhibition and RNA-binding domains of influenza A virus matrix protein mapped with anti-idiotypic antibodies and synthetic peptides. J. Virol. **63:** 3586–3594.
121. WATANABE, K., H. HANDA, K. MIZUMOTO & K. NAGATA. 1996. Mechanism for inhibition of influenza virus RNA polymerase activity by matrix protein. J. Virol. **70:** 241–247.
122. YE, Z., D. ROBINSON & R.R. WAGNER. 1995. Nucleus-targeting domain of the matrix protein (M1) of influenza virus. J. Virol. **69:** 1964–1970.
123. BUI, M., G. WHITTAKER & A. HELENIUS. 1996. Effect of M1 protein and low pH on nuclear transport of influenza virus ribonucleoproteins. J. Virol. **70:** 8391–8401.
124. HUANG, X., T. LIU, J. MULLER, *et al.* 2001. Effect of influenza virus matrix protein and viral RNA on ribonucleoprotein formation and nuclear export. Virology **287:** 405–416.
125. SAKAGUCHI, A., E. HIRAYAMA, A. HIRAKI, *et al.* 2003. Nuclear export of influenza viral ribonucleoprotein is temperature-dependently inhibited by dissociation of viral matrix protein. Virology **306:** 244–253.

126. HIRAYAMA, E., H. ATAGI, A. HIRAKI & J. KIM. 2004. Heat shock protein 70 is related to thermal inhibition of nuclear export of the influenza virus ribonucleoprotein complex. J. Virol. **78:** 1263–1270.
127. GREGORIADES, A. & B. FRANGIONE. 1981. Insertion of influenza M protein into the viral lipid bilayer and localization of site of insertion. J. Virol. **40:** 323–328.
128. BOURMAKINA, S.V. & A. GARCIA-SASTRE. 2003. Reverse genetics studies on the filamentous morphology of influenza A virus. J. Gen. Virol. **84:** 517–527.
129. GOMEZ-PUERTAS, P., C. ALBO, E. PEREZ-PASTRANA, *et al.* 2000. Influenza virus matrix protein is the major driving force in virus budding. J. Virol. **74:** 11538–11547.
130. LIU, T., J. MULLER & Z. YE. 2002. Association of influenza virus matrix protein with ribonucleoproteins may control viral growth and morphology. Virology **304:** 89–96.
131. ARZT, S., I. PETIT, W.P. BURMEISTER, *et al.* 2004. Structure of a knockout mutant of influenza virus M1 protein that has altered activities in membrane binding, oligomerisation and binding to NEP (NS2). Virus Res. **99:** 115–119.
132. HARA, K., M. SHIOTA, H. KIDO, *et al.* 2003. Inhibition of the protease activity of influenza virus RNA polymerase PA subunit by viral matrix protein. Microbiol. Immunol. **47:** 521–526.
133. LAMB, R.A., S.L. ZEBEDEE & C.D. RICHARDSON. 1985. Influenza virus M2 protein is an integral membrane protein expressed on the infected-cell surface. Cell **40:** 627–633.
134. HOLSINGER, L.J. & R.A. LAMB. 1991. Influenza virus M2 integral membrane protein is a homotetramer stabilized by formation of disulfide bonds. Virology **183:** 32–43.
135. SAKAGUCHI, T., Q. TU, L.H. PINTO & R.A. LAMB. 1997. The active oligomeric state of the minimalistic influenza virus M2 ion channel is a tetramer. Proc. Natl. Acad. Sci. USA **94:** 5000–5005.
136. CIAMPOR, F., C.A. THOMPSON, S. GRAMBAS & A.J. HAY. 1992. Regulation of pH by the M2 protein of influenza A viruses. Virus Res. **22:** 247–258.
137. TAKEUCHI, H., A. OKADA & T. MIURA. 2003. Roles of the histidine and tryptophan side chains in the M2 proton channel from influenza A virus. FEBS Lett. **552:** 35–38.
138. TAKEUCHI, K. & R.A. LAMB. 1994. Influenza virus M2 protein ion channel activity stabilizes the native form of fowl plague virus hemagglutinin during intracellular transport. J. Virol. **68:** 911–919.
139. HENKEL, J.R., G. APODACA, Y. ALTSCHULER, *et al.* 1998. Selective perturbation of apical membrane traffic by expression of influenza M2, an acid-activated ion channel, in polarized madin-darby canine kidney cells. Mol. Biol. Cell. **9:** 2477–2490.
140. HENKEL, J.R. & O.A. WEISZ. 1998. Influenza virus M2 protein slows traffic along the secretory pathway. pH perturbation of acidified compartments affects early Golgi transport steps. J. Biol. Chem. **273:** 6518–6524.
141. PARK, E.K., M.R. CASTRUCCI, A. PORTNER & Y. KAWAOKA. 1998. The M2 ectodomain is important for its incorporation into influenza A virions. J. Virol. **72:** 2449–2455.
142. CASTRUCCI, M.R. & Y. KAWAOKA. 1995. Reverse genetics system for generation of an influenza A virus mutant containing a deletion of the carboxyl-terminal residue of M2 protein. J. Virol. **69:** 2725–2728.
143. FIERS, W., M. DE FILETTE, A. BIRKETT, *et al.* 2004. A "universal" human influenza A vaccine. Virus Res. **103:** 173–176.

144. NEIRYNCK, S., T. DEROO, X. SAELENS, *et al.* 1999. A universal influenza A vaccine based on the extracellular domain of the M2 protein. Nat. Med. **5:** 1157–1163.
145. WATANABE, T., S. WATANABE, H. KIDA & Y. KAWAOKA. 2002. Influenza A virus with defective M2 ion channel activity as a live vaccine. Virology **299:** 266–270.
146. LIU, W., H. LI & Y.H. CHEN. 2003. N-terminus of M2 protein could induce antibodies with inhibitory activity against influenza virus replication. FEMS Immunol. Med. Microbiol. **35:** 141–146.
147. LIU, W., P. ZOU & Y.H. CHEN. 2004. Monoclonal antibodies recognizing EVET-PIRN epitope of influenza A virus M2 protein could protect mice from lethal influenza A virus challenge. Immunol. Lett. **93:** 131–136.
148. MOZDZANOWSKA, K., J. FENG, M. EID, *et al.* 2003. Induction of influenza type A virus-specific resistance by immunization of mice with a synthetic multiple antigenic peptide vaccine that contains ectodomains of matrix protein 2. Vaccine **21:** 2616–2626.
149. THOMAS, J.M., M.P. STEVENS, N. PERCY & W.S. BARCLAY. 1998. Phosphorylation of the M2 protein of influenza A virus is not essential for virus viability. Virology **252:** 54–64.
150. WATANABE, T., S. WATANABE, H. ITO, *et al.* 2001. Influenza A virus can undergo multiple cycles of replication without M2 ion channel activity. J. Virol. **75:** 5656–5662.
151. TAKEDA, M., A. PEKOSZ, K. SHUCK, *et al.* 2002. Influenza A virus M2 ion channel activity is essential for efficient replication in tissue culture. J. Virol. **76:** 1391–1399.
152. ALONSO-CAPLEN, F.V. & R.M. KRUG. 1991. Regulation of the extent of splicing of influenza virus NS1 mRNA: role of the rates of splicing and of the nucleocytoplasmic transport of NS1 mRNA. Mol. Cell. Biol. **11:** 1092–1098.
153. NEMEROFF, M.E., U. UTANS, A. KRAMER & R.M. KRUG. 1992. Identification of cis-acting intron and exon regions in influenza virus NS1 mRNA that inhibit splicing and cause the formation of aberrantly sedimenting presplicing complexes. Mol. Cell. Biol. **12:** 962–970.
154. NEMEROFF, M.E., X.Y. QIAN & R.M. KRUG. 1995. The influenza virus NS1 protein forms multimers *in vitro* and *in vivo*. Virology **212:** 422–428.
155. GREENSPAN, D., P. PALESE & M. KRYSTAL. 1988. Two nuclear location signals in the influenza virus NS1 nonstructural protein. J. Virol. **62:** 3020–3026.
156. QIU, Y. & R.M. KRUG. 1994. The influenza virus NS1 protein is a poly(A)-binding protein that inhibits nuclear export of mRNAs containing poly(A). J. Virol. **68:** 2425–2432.
157. HATADA, E., T. TAKIZAWA & R. FUKUDA. 1992. Specific binding of influenza A virus NS1 protein to the virus minus-sense RNA *in vitro*. J. Gen. Virol. **73:** 17–25.
158. MARION, R.M., T. ZURCHER, S. DE LA LUNA & J. ORTIN. 1997b. Influenza virus NS1 protein interacts with viral transcription-replication complexes *in vivo*. J. Gen. Virol. **78:** 2447–2451.
159. HATADA, E. & R. FUKUDA. 1992. Binding of influenza A virus NS1 protein to dsRNA *in vitro*. J. Gen. Virol. **73:** 3325–3329.
160. QIU, Y., M. NEMEROFF & R.M. KRUG. 1995. The influenza virus NS1 protein binds to a specific region in human U6 snRNA and inhibits U6-U2 and U6-U4 snRNA interactions during splicing. RNA **1:** 304–316.

161. FORTES, P., A. BELOSO & J. ORTIN. 1994. Influenza virus NS1 protein inhibits pre-mRNA splicing and blocks mRNA nucleocytoplasmic transport. EMBO J. **13:** 704–712.
162. LU, Y., X.Y. QIAN & R.M. KRUG. 1994. The influenza virus NS1 protein: a novel inhibitor of pre-mRNA splicing. Genes Dev. **8:** 1817–1828.
163. ALONSO-CAPLEN, F.V., M.E. NEMEROFF, Y. QIU & R.M. KRUG. 1992. Nucleocytoplasmic transport: the influenza virus NS1 protein regulates the transport of spliced NS2 mRNA and its precursor NS1 mRNA. Genes Dev. **6:** 255–267.
164. MARION, R.M., T. ARAGON, A. BELOSO, *et al.* 1997a. The N-terminal half of the influenza virus NS1 protein is sufficient for nuclear retention of mRNA and enhancement of viral mRNA translation. Nucleic Acids Res. **25:** 4271–4277.
165. QIAN, X.Y., F. ALONSO-CAPLEN & R.M. KRUG. 1994. Two functional domains of the influenza virus NS1 protein are required for regulation of nuclear export of mRNA. J. Virol. **68:** 2433–2441.
166. CHEN, Z., Y. LI & R.M. KRUG. 1999. Influenza A virus NS1 protein targets poly(A)-binding protein II of the cellular 3′-end processing machinery. EMBO J. **18:** 2273–2283.
167. NEMEROFF, M.E., S.M. BARABINO, Y. LI, *et al.* 1998. Influenza virus NS1 protein interacts with the cellular 30 kDa subunit of CPSF and inhibits 3′end formation of cellular pre-mRNAs. Mol. Cell. **1:** 991–1000.
168. LU, Y., M. WAMBACH, M.G. KATZE & R.M. KRUG. 1995. Binding of the influenza virus NS1 protein to double-stranded RNA inhibits the activation of the protein kinase that phosphorylates the elF-2 translation initiation factor. Virology **214:** 222–228.
169. DE LA LUNA, S., P. FORTES, A. BELOSO & J. ORTIN. 1995. Influenza virus NS1 protein enhances the rate of translation initiation of viral mRNAs. J. Virol. **69:** 2427–2433.
170. ENAMI, K., T.A. SATO, S. NAKADA & M. ENAMI. 1994. Influenza virus NS1 protein stimulates translation of the M1 protein. J. Virol. **68:** 1432–1437.
171. GARCIA-SASTRE, A., A. EGOROV, D. MATASSOV, *et al.* 1998. Influenza A virus lacking the NS1 gene replicates in interferon-deficient systems. Virology **252:** 324–330.
172. CHEUNG, C.Y., L.L. POON, A.S. LAU, *et al.* 2002. Induction of proinflammatory cytokines in human macrophages by influenza A (H5N1) viruses: a mechanism for the unusual severity of human disease? Lancet **360:** 1831–1837.
173. ODAGIRI, T., K. TOMINAGA, K. TOBITA & S. OHTA. 1994. An amino acid change in the non-structural NS2 protein of an influenza A virus mutant is responsible for the generation of defective interfering (DI) particles by amplifying DI RNAs and suppressing complementary RNA synthesis. J. Gen. Virol. **75:** 43–53.
174. WARD, A.C., L.A. CASTELLI, A.C. LUCANTONI, *et al.* 1995. Expression and analysis of the NS2 protein of influenza A virus. Arch. Virol. **140:** 2067–2073.
175. NEUMANN, G., M.T. HUGHES & Y. KAWAOKA. 2000. Influenza A virus NS2 protein mediates vRNP nuclear export through NES-independent interaction with hCRM1. EMBO J. **19:** 6751–6758.
176. HERZ, C., E. STAVNEZER, R. KRUG & T. GURNEY, JR. 1981. Influenza virus, an RNA virus, synthesizes its messenger RNA in the nucleus of infected cells. Cell **26:** 391–400.

177. JACKSON, D.A., A.J. CATON, S.J. MCCREADY & P.R. COOK. 1982. Influenza virus RNA is synthesized at fixed sites in the nucleus. Nature **296:** 366–368.
178. ROBERTSON, J.S. 1979. 5′ and 3′ terminal nucleotide sequences of the RNA genome segments of influenza virus. Nucleic Acids Res. **6:** 3745–3757.
179. SKEHEL, J.J. & A.J. HAY. 1978. Nucleotide sequences at the 5′ termini of influenza virus RNAs and their transcripts. Nucleic Acids Res. **5:** 1207–1219.
180. HAY, A.J., G. ABRAHAM, J.J. SKEHEL, *et al.* 1977a. Influenza virus messenger RNAs are incomplete transcripts of the genome RNAs. Nucleic Acids Res. **4:** 4197–4209.
181. ROBERTSON, J.S., M. SCHUBERT & R.A. LAZZARINI. 1981. Polyadenylation sites for influenza virus mRNA. J. Virol. **38:** 157–163.
182. HAY, A.J., B. LOMNICZI, A.R. BELLAMY & J.J. SKEHEL. 1977b. Transcription of the influenza virus genome. Virology **83:** 337–355.
183. HAY, A.J., J.J. SKEHEL & J. MCCAULEY. 1982. Characterization of influenza virus RNA complete transcripts. Virology **116:** 517–522.
184. SHAPIRO, G.I., T. GURNEY, JR. & R.M. KRUG. 1987. Influenza virus gene expression: control mechanisms at early and late times of infection and nuclear-cytoplasmic transport of virus-specific RNAs. J. Virol. **61:** 764–773.
185. HAY, A.J. & J.J. SKEHEL. 1979. Characterization of influenza virus RNA transcripts synthesized *in vitro*. J. Gen. Virol. **44:** 599–608.
186. KAWAKAMI, K., A. ISHIHAMA & M. HAMAGUCHI. 1981. RNA polymerase of influenza virus. I. Comparison of the virion-associated RNA polymerase activity of various strains of influenza virus. J. Biochem. **89:** 1751–1757.
187. KRUG, R.M., B.A. BRONI, A.J. LAFIANDRA, *et al.* 1980. Priming and inhibitory activities of RNAs for the influenza viral transcriptase do not require base pairing with the virion template RNA. Proc. Natl. Acad. Sci. USA **77:** 5874–5878.
188. SHIH, S.R. & R.M. KRUG. 1996. Surprising function of the three influenza viral polymerase proteins: selective protection of viral mRNAs against the cap-snatching reaction catalyzed by the same polymerase proteins. Virology **226:** 430–435.
189. FLICK, R., G. NEUMANN, E. HOFFMANN, *et al.* 1996. Promoter elements in the influenza vRNA terminal structure. RNA **2:** 1046–1057.
190. LUO, G.X., W. LUYTJES, M. ENAMI & P. PALESE. 1991. The polyadenylation signal of influenza virus RNA involves a stretch of uridines followed by the RNA duplex of the panhandle structure. J. Virol. **65:** 2861–2867.
191. CHEONG, H.K., C. CHEONG & B.S. CHOI. 1996. Secondary structure of the panhandle RNA of influenza virus A studied by NMR spectroscopy. Nucleic Acids Res. **24:** 4197–4201.
192. BEATON, A.R. & R.M. KRUG. 1981. Selected host cell capped RNA fragments prime influenza viral RNA transcription *in vivo*. Nucleic Acids Res. **9:** 4423–4436.
193. HAGEN, M., L. TILEY, T.D. CHUNG & M. KRYSTAL. 1995. The role of template-primer interactions in cleavage and initiation by the influenza virus polymerase. J. Gen. Virol. **76:** 603–611.
194. POON, L.L., D.C. PRITLOVE, J. SHARPS & G.G. BROWNLEE. 1998. The RNA polymerase of influenza virus, bound to the 5′ end of virion RNA, acts in cis to polyadenylate mRNA. J. Virol. **72:** 8214–8219.

195. Plotch, S.J. & R.M. Krug. 1977. Influenza virion transcriptase: synthesis *in vitro* of large, polyadenylic acid-containing complementary RNA. J. Virol. **21:** 24–34.
196. Poon, L.L., D.C. Pritlove, E. Fodor & G.G. Brownlee. 1999. Direct evidence that the poly(A) tail of influenza A virus mRNA is synthesized by reiterative copying of a U track in the virion RNA template. J. Virol. **73:** 3473–3476.
197. Tiley, L.S., M. Hagen, J.T. Matthews & M. Krystal. 1994. Sequence-specific binding of the influenza virus RNA polymerase to sequences located at the 5′ ends of the viral RNAs. J. Virol. **68:** 5108–5116.
198. Li, X. & P. Palese. 1994. Characterization of the polyadenylation signal of influenza virus RNA. J. Virol. **68:** 1245–1249.
199. Poon, L.L., E. Fodor & G.G. Brownlee. 2000. Polyuridylated mRNA synthesized by a recombinant influenza virus is defective in nuclear export. J. Virol. **74:** 418–427.
200. Eckner, R., W. Ellmeier & M.L. Birnstiel. 1991. Mature mRNA 3′ end formation stimulates RNA export from the nucleus. EMBO J. **10:** 3513–3522.
201. Hamm, J. & I.W. Mattaj. 1990. Monomethylated cap structures facilitate RNA export from the nucleus. Cell **63:** 109–118.
202. Beelman, C.A., A. Stevens, G. Caponigro, *et al.* 1996. An essential component of the decapping enzyme required for normal rates of mRNA turnover. Nature **382:** 642–646.
203. Carpousis, A.J., N.F. Vanzo & L.C. Raynal. 1999. mRNA degradation. A tale of poly(A) and multiprotein machines. Trends Genet. **15:** 24–28.
204. Nagata, K., K. Takeuchi & A. Ishihama. 1989. *In vitro* synthesis of influenza viral RNA: biochemical complementation assay of factors required for influenza virus replication. J. Biochem. **106:** 205–208.
205. Shimizu, K., H. Handa, S. Nakada & K. Nagata. 1994. Regulation of influenza virus RNA polymerase activity by cellular and viral factors. Nucleic Acids Res. **22:** 5047–5053.
206. Smith, G.L. & A.J. Hay. 1982. Replication of the influenza virus genome. Virology **118:** 96–108.

The Molecular Biology of SARS Coronavirus

NAMITA SATIJA AND SUNIL K. LAL

Virology Group, International Centre for Genetic Engineering & Biotechnology, New Delhi, India

ABSTRACT: Severe acute respiratory syndrome (SARS) is the first emerging infectious disease of the 21st century that has been highly transmissible and fatal and was caused by a previously unknown coronavirus (SARS-CoV). The SARS epidemic in 2003 resulted in more than 8400 SARS cases and approximately 800 deaths. Existing in non-identified animal reservoirs, SARS-CoV continues to represent a threat to humans although more than four years have passed since a large outbreak of SARS, and no new cases have been reported. However, we cannot exclude the possibility of reemergence of SARS. It is hence necessary to understand the biology of the SARS-CoV to deal adequately with the next outbreak, whenever it happens. The SARS-CoV is a novel coronavirus with a large (∼30 thousand nucleotides) positive-sense, single-stranded RNA containing 14 functional open reading frames (ORFs) of which 2 large ORFs constitute the replicase gene which encodes proteins required for viral RNA syntheses. The remaining 12 ORFs encode the 4 structural proteins: spike, membrane, nucleocapsid and envelope; and eight accessory proteins. The viral genome and its expression within the host cell undergoes extensive translational and enzymatic processing to form the 4 structural, 8 accessory and 16 nonstructural proteins. In an effort to understand the molecular mechanisms or capsid assembly and viral pathogenesis, laboratories around the world have adopted a variety of approaches to answering these trivial questions. It has been our effort to consolidate all information known to date about the molecular mechanisms of the SARS-CoV into this chapter to update our readership on the current status of research.

KEYWORDS: coronavirus; structural proteins; accessory proteins; viral capsid; assembly; proteolytic cleavage

Address for correspondence: Sunil K. Lal, Virology Group, International Centre for Genetic Engineering & Biotechnology, P.O. Box 10504, Aruna Asaf Ali Road, New Delhi 110067, India. Voice: 91-11-26177357; fax: 91-11-26162316.
sunillal@icgeb.res.in

Ann. N.Y. Acad. Sci. 1102: 26–38 (2007). © 2007 New York Academy of Sciences.
doi: 10.1196/annals.1408.002

INTRODUCTION

Severe acute respiratory syndrome (SARS) is a newly emerged infectious disease that appeared in Guangdong Province, mainland China, in November 2002. By March 2003, the disease had spread globally and by July there were 8,447 probable SARS cases including 811 deaths reported from 32 countries or regions worldwide to the WHO.

The etiological agent was identified as a coronavirus (CoV)[1] and the complete genome sequence of SARS-CoV Tor2 (Toronto) isolate was published in April 2003, which established it as a novel member of the family.[2,3]

The genome of SARS-CoV is a positive sense single-stranded ribonucleic acid (RNA) consisting of 29,751 bases. Phylogenetic analysis revealed that this CoV was only moderately related to other known CoVs, including two human CoVs, HCoV-OC43 and HCoV-229E, known to cause common colds and lower respiratory tract infections and diarrhea in humans.

The clinical symptoms of the syndrome include fever, chills, rigors, cough, and headache and the pathological aspects include lymphopenia, thrombocytopenia, elevated lactate dehydrogenase, and creatine kinase levels.

Although currently the spread of the virus seems to be confined to rigorous and timely quarantine measures (http://www.who.int/csr/sars/country/table2003 0923/en/), it may still be circulating in the animal reservoir and it is impossible to predict when it will return. Because this scare of recurrence of SARS looms over mankind, better monitoring of SARS outbreaks through accurate diagnostic tests and the development of effective antiviral therapies are urgently required. These in turn depend directly on better molecular understanding of the SARS-CoV infection cycle. Research toward the understanding of the SARS virus at a molecular level is therefore of vital importance to combat SARS-CoV infection if another SARS pandemic was to occur in the future.

TISSUE TROPISM

SARS mainly shows pneumonia-like symptoms and the lung is pathologically the most affected organ. Detailed investigations, however, suggest that SARS is a systemic disease with widespread extrapulmonary dissemination, resulting in viral shedding in respiratory secretions, stools, urine, and possibly even in sweat.[4,5] Immunohistochemistry and *in situ* hybridization of organs from four SARS patients who died revealed that the virus was found not only in the lung and intestine, but also in the liver, distal convoluted renal tubules, sweat glands, parathyroid, pituitary, pancreas, adrenal gland, and cerebrum.[4] SARS-CoV RNA was detected in the lung, small and large bowels, lymph nodes, spleen, liver, heart, kidney, and skeletal muscle, in descending order of viral load per gram of tissue using reverse-transcriptase polymerase chain reaction.[5]

CELLULAR RECEPTOR

The virus seems to gain entry into the target cells by direct membrane fusion at the target cell surface. Apart from this it has been postulated that the SARS-CoV may gain cell entry via pH-dependent endocytosis. The spike protein of the virus is the key player in this regard.[6] A metallopeptidase, angiotensin-converting enzyme 2 (ACE2), the C-type lectin CD209L (also known as L-SIGN), and DC-SIGN bind SARS-CoV, but ACE2 appears to be the key functional receptor for the virus.

The ACE2 protein is reportedly present in type 1 and type 2 pneumocytes, enterocytes of all parts of the small intestine, the brush border of the proximal tubular cells of the kidney as well as the endothelial cells of small and large arteries and veins of all tissues studied, and arterial smooth muscle cells.[7] This localization of ACE2 explains the tissue tropism of SARS-CoV for the lung, small intestine, and kidney; however, notable discrepancies include virus replication in colonic epithelium, which has no ACE2, and no virus infection in endothelial cells, which have ACE2. Other receptors or co-receptors such as L-SIGN may explain such discrepancies.

GENOME ORGANIZATION AND PROTEIN PROCESSING

The SARS-CoV genome has been predicted to contain 14 functional open reading frames (ORFs). The comparison of different SARS-CoV ORFs with those of other CoVs revealed a similar pattern of structural gene arrangement with the replicase and protease genes (gene 1a-1b) and the spike (S), envelope (E), membrane (M), and nucleocapsid (N) genes in a 5′ to 3′ order of appearance. Interspersed between these well-characterized genes is a series of ORFs, many of whose functions are yet unknown. There are two ORFs situated between the S and the E genes and three to five ORFs between the M and N genes (FIG.1). Such gene organization most closely resembles that of group III CoVs. Also, the SARS-CoV genomic sequence does not contain a gene for the hemagglutinin-esterase (HE) protein, which is present in most group II CoVs.

Two-thirds of the SARS-CoV RNA is organized in the genes 1a-1b. The sequence of these genes is highly conserved among all CoVs with the exception that SARS-CoV lacks the sequence coding for papain-like protease (PL1pro), one of the two papain-like proteinases operating on cleavage sites at the N terminus of the polyproteins. ORFs 1a and 1b encode two polyproteins, pp1a and pp1ab, the latter is predicted to translate through a -1 ribosomal frameshifting mechanism. These two polyproteins are processed by virus-encoded proteinases, to yield 16 individual nonstructural proteins (nsp). Many of the postulated nsp proteins to date have undesignated functions but it is suggested that they participate primarily in viral RNA replication. Most potential gene 1a-1b products are fairly well conserved between SARS-CoV and other CoVs. PL2pro of the SARS virus is responsible for the cleavage of all the N-terminal proteins of gene 1a. The main proteinase (Mpro), also called the 3C-like protease (3CLpro), is responsible for the cleavage of all the remaining proteins encoded by gene 1a-1b.

The SARS-CoV 5′ and 3′ UTR

The 5′ UTR (untranslated region) of the SARS-CoV genome was initially characterized by 5′ rapid amplification of the cDNA ends and Northern blot assays. These procedures elucidated the leader sequence and the transcription regulatory sequence (TRS). The leader sequences found in the viral sg-mRNA (subgenomic-mRNA) transcripts were at least 72 nucleotides long. After aligning the leader sequence at the 5′ end of the eight sg-mRNAs, a minimal consensus TRS was found with the sequence 5′-ACGAAC-3′,[8] which participated in the discontinuous synthesis of sg-mRNAs as a signaling sequence. TRS showed no clear relationship with the abundance of the sg-mRNAs. A highly conserved s2m motif with 32 nucleotides was also identified in the 3′ region of the genome, which has also been discovered in the avian infectious bronchitis virus.

The Replicase Gene

The replicase gene of the SARS-CoV encodes for two proteins as a consequence of the proteolytic processing of the large polyproteins (pp1a and pp1ab). The translation of segment 1b of such a polyprotein is interrupted by the -1 ribosomal frameshifting by a putative "slippery" sequence and a corresponding putative pseudoknot structure, which is required for stalling the translating ribosome at the slippery site to undergo -1 translational frameshifting. Two functional domains, the papain-like cysteine proteinase (PL2pro) and the 3C-like cysteine proteinase (3CLpro), were identified experimentally and are considered responsible for the proteolytic processing of the polyprotein into 16 subunits. A 375 amino acid SARS-CoV unique domain was identified upstream of the PL2pro domain, not found in any other known CoVs. In addition, seven more putative regions encoding RNA processing enzymes were identified, namely, RNA-dependent RNA polymerase (RDRP), RNA helicase (HEL), poly (U)-specific endoribonuclease (NendoU), 3′-5′ exonuclease (ExoN), S-adenosylmethionine-dependent ribose 2′-O-methyltransferase (2′-O-MT), adenosine diphosphate-ribose 1′-phosphatase (ADRP), and a cyclic phosphodiesterase (CPD). The translation of two polyproteins from the ORF 1a and 1b starts the genome expression (FIG. 1). The polyproteins are then proteolytically processed by PL2pro and 3CLpro into 16 units. PL2pro is responsible for the N-proximal cleavage and 3CLpro is responsible for the

C-proximal cleavage. The helicase is then released. ATPase activity and DNA duplex-unwinding activity were demonstrated by purified helicase, indicating that the protein has RNA polymerase activity.[9]

The Spike Protein

Together with the M protein, the spike (S) protein is believed to be incorporated into the viral envelope before the mature virion is released. Initial analysis of the 1255a.a. peplomer protein of the virus reveals the possible existence of a signal peptide that would most likely be cleaved between residues 13 and 14.[11,12] It forms typical petal-shaped spikes on the surface of the virion and consists of three domains, the external N-terminal domain with its conserved S1 and S2 subdomains, a transmembrane domain, and a short cytoplasmic domain at the C terminus. S1 is the receptor-binding unit in the N terminus. Molecular modeling of the S1 and S2 subunits of the S glycoprotein suggest that the former unit consists mainly of antiparallel β-sheets with dispersed α and β regions, in addition to the three domains identified in the S2 unit. Entry of CoVs into target cells is initiated by binding of the viral S protein to receptor molecules. The S1 domain of SARS-CoV is significantly different from that of other human CoVs, implying that these viruses are using different receptors for host cell entry. The cellular receptor of several group I CoVs are aminopeptidase N and zinc metalloprotease, whereas mouse hepatitis virus (MHV) (group II) uses carcinoembryonic antigen-related cell adhesion molecules as a cellular receptor. In the case of SARS-CoV, ACE2 was demonstrated to be a functional receptor for the SARS-CoV *in-vitro*. Syncytia were observed in cell culture expressing ACE2 and the SARS-CoV S1 domain, which could be inhibited using anti-ACE2 antibody. Fine mapping on the N-terminal unit of the S protein indicates that the minimal receptor-binding domain is located between the residues 318–510.[10] Antibodies specific for this S1 subunit of the SARS-CoV S protein were shown to neutralize SARS-CoV infection. The C-type lectin CD209L (also called L-SIGN), a human cellular glycoprotein has been reported to serve as an alternative receptor for SARS-CoV. The S protein of SARS-CoV contains 23 putative N-glycosylation sites, among which 12 have been described to be effectively glycosylated.[11,12] The S protein, when expressed alone, acquires EndoH-resistant complex N-glycans in the Golgi within 30 min following expression. Both high-mannose and complex glycan N-glycoforms have been detected on S trimers within ER and Golgi, respectively, suggesting that trimers form in the ER and pass the quality control to move toward the Golgi to acquire complex N-glycans. It has also been shown that the S glycoprotein is present all along the secretory pathway from the ER to the plasma membrane.[13]

Recently, structure of SARS spike protein's receptor-binding domain in complex with its receptor, ACE2 has been determined using X-ray crystallography. The structure though determined at 2.9A, shows potential for usage in designing vaccines against the virus.[14]

The Nucleocapsid Protein

Nucleocapsid or the N protein of SARS-CoV is a 46-kDa phosphoprotein, which shows homology with other members of CoV family. Similar to other CoVs, it gets abundantly expressed during infection and antibodies against N have been detected in SARS patients.[15–18] N protein of most CoVs contains multiple epitopic sites lying most commonly at their C terminus. The epitope site for TGEV is located around codons 360–382,[19] for MHV it is at 381–405[20] and for infectious bronchitis virus (IBV) it is located at 360–409,[21] all at the C terminus. Likewise SARS-CoV, N protein has been found to contain its most antigenic site at the C terminus from codons 371–407.[22] This makes N an excellent candidate for diagnostic purposes.

Being a multipurpose protein, it plays a vital role during virion assembly. The process of virion assembly requires efficient packaging of the viral RNA into the virion. For this, N protein associates with the viral genomic RNA, and together they form the ribonucleoprotein. In MHV it has been shown that N protein does possess RNA-binding activity.[16] The structure of the N terminal (49–178 aa) region of the N protein has been solved by a group at the Abbott Laboratories, Illinois, USA. The structure reveals an RNA-binding domain similar to U1A RNA-binding protein.[23] Studies have also revealed that packaging of SARS RNA into virus-like particles (VLPs) is dependent on N protein. Packaging signal within a domain from nucleotide 19715–20294 of genomic RNA and the N terminal structure of N protein are essential for packaging and RNA-binding activity, respectively.[24] Purified SARS-CoV has been shown to have RNA chaperone activity.[51]

Self-association of N protein governs the formation of viral capsid to a large extent and has been observed in many viruses. For SARS-CoV, two groups have shown that N protein does dimerize. The interaction domain responsible for dimerization was first shown to be residing in the C terminal 209 amino acid residues[25] and was further narrowed down to 138 amino acid residues (between 285 and 422) at the C terminus of the protein.[26] N's ability to form oligomers (trimers, tetramers, and hexamers) has been demonstrated to be residing in residues 343–402.[49] N also interacts with the membrane protein. Amino acids 211–254, especially tetrad glutamines (^{240}QQQQ^{243}), are essential for this interaction.[50]

Apart from interactions among viral proteins, interactions with the proteins of the host play a major role in the establishment of viral infection. Hence mapping these viral host protein interactions provides vital hints about how infection actually takes place and progresses. It has been shown that N binds tightly to human cyclophilin A (cypA).[27] CypA is a family of proteins that has been extensively studied but their function remains unknown. Recently, cyclophilins have been shown to play an important role in HIV infection. CypA specifically has been shown to play a vital role in viral core disassembly.[28] Hence CypA interaction with N protein may provide new hints about how SARS CoV infects host cells.

N protein has also been reported to bind to human cellular heterogenous nuclear ribonucleoprotein A1 (hnRNP A1) with high affinity.[29] The regions responsible for this protein–protein interaction have been shown to be residing in the region from amino acid 161–210 of SARS CoV and C terminus amino acid 203–320 of human hnRNP A1. Studies have indicated that hnRNP A1 acts as a global regulator of alternative pre-mRNA splicing and its activities result in inactivation of alternative distal 5′ splicing sites and skipping of optional exons.[30] Also a recent publication has demonstrated that the nucleocapsid protein of MHV could interact with hnRNP A1.[31] Hence it may be postulated that interaction of SARS-CoV, N protein with hnRNP A1 may play a role in transcription and replication of viral genome.

Another aspect important for establishing an infection is modulation of host pathways by viral proteins. A report by He *et al.* in 2003 has shown that N protein activates AP1 signal transduction pathway. However, the mechanism underlying the observation is not known yet. Neither is it known whether N interacts with the components of AP1 directly or through an intermediate. AP1 or activator protein 1 pathway is a regulator of a wide variety of cellular processes. These include cell proliferation, differentiation, and apoptosis. This makes this pathway an important target for modification by the viral proteins to gain control over the host cell cycle.[32] Moreover, binding of N protein to cyclin-dependent kinase (CDK) complex leads to downregulation of S phase gene products, which in turn leads to inhibition of S phase progression.[33]

SARS-CoV N protein has also been shown to induce apoptosis in Cos 1 cells. Cells transfected with N protein show significant cell death under conditions of stress, induced by serum starvation. N protein has been found to downregulate extracellular signal-regulated kinase (ERK), but upregulate c-jun-N-terminal kinase (JNK) and p38 mitogen-activated map kinase (MAPK). Further, p38 MAPK activation was also found to induce actin reorganization.[34]

Nucleocapsid protein of several CoVs exhibits cytoplasmic as well as nucleolar localization at some point of time during the infection cycle. In the case of SARS-CoV, the N protein had been predicted to contain as many as eight nuclear localization signal (NLS) motifs, both monopartite and bipartite. A report by Zhu *et al.*[35] showed that the wild-type N was distributed mainly in the cytoplasm. The N-terminal of N, which contains the NLS1 (aa38–44), was localized to the nucleus. The C terminus of N, which contains both NLS2 (aa257–265) and NLS3 (aa369–390), was localized to the cytoplasm and the nucleolus. Also the region containing amino acids 226–289 was able to mediate nucleolar localization. Furthermore, deletion of the leucine-rich region (220-LALLLLDRLNRL) resulted in the accumulation of N to the cytoplasm and nucleolus, and when fusing this peptide to enhanced green fluorescence protein (EGFP) localization was cytoplasmic, suggesting that N may act as a shuttle protein.[35] A report by Surjit *et al.* (2005) provides evidence that N is indeed a nucleocytoplasmic-shuttling protein.[36] N proteins of many CoVs have been shown to be phosphorylated.[37] However, the mechanism by which this happens and the functional significance of this process remains largely unknown. N protein of SARS-CoV has been shown to be a phosphoprotein, which is stable and localizes in the cytoplasm. The phosphorylation takes place at the serine residues and that too by multiple kinases. It has been shown to be a substrate for CDK, glycogen synthase kinase, MAPK, and casein kinase II. A possible mechanism for phosphorylation-dependent nucleocytoplasmic shuttling of the N protein has also been elucidated. It has been shown that phosphorylated N is translocated to cytoplasm with the help of 14-3-3 (tyrosine 3-monooxygenase/tryptophan 5-monooxygenase activation protein), a protein family that functions as adapters for modulating interactions between components of various cellular signaling and cell-cycle regulatory pathways through phosphorylation-dependent protein–protein interactions.[36]

It can be postulated that N phosphorylation may be serving an array of purposes ranging from acting as a deciding factor for localization of protein to aiding in creating a more favorable environment for replication of the virus.[36]

The Matrix Protein

CoV matrix (M) protein is the most abundant structural protein at the surface of virus particles and contains three putative transmembrane domains, a short N-terminal ectodomain, and a long C-terminal endodomain.

Interaction of the M protein with N has been described for many CoVs. For SARS-CoV, the M–N interaction was first reported by Fang *et al.* (2005) using glutathione-*S*-transferase pulldowns. Laboratory experiments have shown that the 12 amino acids domain (194–205) in the M protein is responsible for binding to the N protein.[39]

Linear epitope mapping of the M protein using synthetic peptides revealed that amino acid residues 2,137,158 interacted with SARS patient sera by enzyme-linked immunosorbent assay, implying the potential capability of the M protein to induce an immune response. The M protein resides on the viral envelope and in the internal core. In the cells, the M protein is anchored in the Golgi complex, thus dictating the site of virus assembly to the ER-Golgi complex.

It has been shown that SARS-CoV M protein is N-glycosylated at Asn 4 residue in mammalian cells.[40] Also C-terminally tagged M glycoprotein strongly co-localizes with Golgi markers and partially with the ERGIC-53 protein, a lectin that cycles between ER, ERGIC, and cis-Golgi.[41]

The Envelope Protein

The envelope (E) protein of the SARS-CoV is, as the name suggests, the main component of the virus envelope. It is an integral membrane protein,

76 amino acids long, and is highly hydrophobic in nature. In mammalian cells it is found to be localized to endoplasmic reticulum (ER), Golgi apparatus, and lipid rafts of cell membrane. The transmembrane domain of the E protein has been found to alter the membrane permeability.[42]

The ORF 3a or X1

ORF 3a is also called SARS X1 and was initially named U274. The 3a locus encodes an ORF of unknown function and shows no homology to any known protein. The 3a mRNA and 3a protein have been detected in SARS-CoV-infected monkey kidney Vero E6 cells. In addition, the 3a protein has been detected in the lung tissue of a SARS patient. A signal sequence and three transmembrane domains have been predicted when 3a sequence was searched against SMART server.[43] It is a novel SARS-CoV protein that has been shown to be transported efficiently to the cell surface of infected and transfected cells. Its cytoplasmic domain contains sorting signals important for this process. The 3a protein expressed on the cell surface can also undergo endocytosis. It has been shown to interact with M, E, and S proteins suggesting that it may play a role in viral assembly and/or release of virus from infected cells.[44] Also it has been shown to form ion channels in Xenopus oocytes. Formation of a pore in the infected cells may aid in virus release. The modulation of virus release may also be one of the important functions of 3a as reduction in 3a protein expression in SARS-infected FRhK4 cells led to a decrease in virus release.[48]

3a has also been shown to upregulate expression of fibrinogen in lung epithelial cells. mRNA levels of all three subunits of fibrinogen Aα, Bβ, and γ were found to be upregulated. As a result of this intracellular levels as well as secretion of fibrinogen were found to increase.[45]

In addition, 3a protein has also been shown to be expressed in the lungs and intestinal tissues of SARS patients and that the protein localized to the ER in 3a-transfected monkey kidney Vero E6 cells. *In vitro* experiments of chromatin condensation and DNA fragmentation suggested that the 3a protein may trigger apoptosis through a caspase-8-dependent pathway in Vero E6 cells.[46]

The ORF 7a or X4

Like ORF 3a, sequence homology search yielded no significant result for any existing proteins, but the existence of a cleavage site (between residues 15 and 16) and a transmembrane helix were predicted for the 7a protein. It has recently been shown that overexpression of 7a induces apoptosis via a caspase-dependent pathway in different cell lines derived from the lung, kidney, and liver.[47]

CONCLUSIONS

Four years since the discovery of SARS, a great deal of scientific coordination across the globe has led to unfolding of various facts about this virus. But much needs to be learned about pathogenesis of the disease and biology of the virus. These aspects are not fully understood as yet and consorted efforts and collaboration among the scientific community are required to unravel the mysteries surrounding the first pandemic of the 21st century. Reemergence of SARS looms as an uncertain possibility.

The challenge now is to maintain interest in efforts to understand the biology of the virus and pathogenesis of the disease caused by it so as to minimize the risk of being surprised by its sudden reappearance. The lessons learned from SARS would certainly be helpful to counter any reemergence of SARS itself or emergence of any other pandemic.

REFERENCES

1. Peiris, J.S., C.M. Chu, V.C. Cheng, *et al.* HKU/UCH SARS Study Group. 2003. Coronavirus as a possible cause of severe acute respiratory syndrome. Lancet **361:** 1319–1325.
2. Rota, P.A., M.S. Oberste, S.S. Monroe, *et al.* 2003. Characterization of a novel coronavirus associated with severe acute respiratory syndrome. Science **300:** 1394–1399.
3. Marra, M.A., S.J. Jones, C.R. Astell, *et al.* 2003. The genome sequence of the SARS-associated coronavirus. Science **300:** 1399–1404.
4. Ding, Y., L. He, Q. Zhang, *et al.* 2004. Organ distribution of severe acute respiratory syndrome (SARS) associated coronavirus (SARS-CoV) in SARS patients: implications for pathogenesis and virus transmission pathways. J. Pathol. **203:** 622–630.
5. Farcas, G.A., S.M. Poutanen, T. Mazzulli, *et al.* 2005. Fatal severe acute respiratory syndrome is associated with multiorgan involvement by coronavirus. J. Infect. Dis. **191:** 193–197.
6. Yang, Z.Y., Y. Huang, L. Ganesh, *et al.* 2004. pH-dependent entry of severe acute respiratory syndrome coronavirus is mediated by the spike glycoprotein and enhanced by dendritic cell transfer through DC-SIGN. J. Virol. **78:** 5642–5650.
7. Hamming, I., W. Timens, M.L.C. Bulthuis, *et al.* 2004. Tissue distribution of ACE2 protein, the functional receptor for SARS coronavirus. A first step in understanding SARS pathogenesis. J. Pathol. **203:** 631–637.
8. Theil, V., K.A. Ivanov, A. Putics, *et al.* 2003. Mechanisms and enzymes involved in SARS coronavirus genome expression. J. Gen. Virol. **84:** 2305–2315.
9. Zeibuhr, J. 2004. Molecular biology of severe acute respiratory syndrome coronavirus. Curr. Opin. Microbiol. **7:** 412–419.
10. Wong, S.K., L. Wenhui, M.J. Moore, *et al.* 2004. A 193 amino acid fragment of SARS coronavirus S protein efficiently binds angiotensin converting enzyme 2. J. Biol. Chem. **279:** 3197–3201.

11. KROKHIN, O., Y. LI, A. ANDONOV, *et al.* 2003. Mass spectrometric characterization of proteins from the SARS virus: a preliminary report. Mol. Cell Prot. **2:** 346–356.
12. YING, W., Y. HAO, Y. ZHANG, *et al.* 2004. Proteomic analysis on structural proteins of severe acute respiratory syndrome coronavirus. Proteomics **4:** 492–504.
13. NAL, B., C. CHAN, F. KIEN, *et al.* 2005. Differential maturation and subcellular localization of severe acute respiratory syndrome coronavirus surface proteins S, M and E. J. Gen. Virol. **86:** 1423–1434.
14. LI, F., W. LI, M. FARZAN & S.C. HARRISON. 2005. Structure of SARS coronavirus spike receptor-binding domain complexed with receptor. Science **309:** 1864–1868.
15. LIU, X., Y. SHI, P. LI, *et al.* 2004. Profile of antibodies to the nucleocapsid protein of the severe acute respiratory syndrome (SARS)-associated coronavirus in probable SARS patients. Clin. Diagn. Lab. Immunol. **11:** 227–228.
16. SHI, Y., Y. YI, P. LI, *et al.* 2003. Diagnosis of severe acute respiratory syndrome (SARS) by detection of SARS coronavirus nucleocapsid antibodies in an antigen-capturing enzyme-linked immunosorbent assay. J. Clin. Microbiol. **41:** 5781–5782.
17. WANG, J., J. WEN, J. LI, *et al.* 2003. Assessment of immunoreactive synthetic peptides from the structural proteins of severe acute respiratory syndrome coronavirus. Clin. Chem. **49:** 1989–1996.
18. YUXIAN, H., Y. ZHOU, H. WU, *et al.* 2004. Mapping of antigenic sites on the nucleocapsid protein of the severe acute respiratory syndrome coronavirus. J. Clin. Microbiol. **42:** 5309–5314.
19. MARTIN, A.J.M. 1992. Antigenic structure of transmissible gastroenteritis virus nucleoprotein. Virology **188:** 168–174.
20. STOHLMAN, S.A. 1994. Location of antibody epitopes within the mouse hepatitis virus nucleocapsid protein. Virology **202:** 146–153.
21. SEAH, J.N. 2000. Localization of linear B-cell epitopes on infectious bronchitis virus nucleocapsid protein. Vet. Microbiol. **75:** 11–16.
22. LI, S., L. LIN, H. WANG, *et al.* 2003. The epitope study on the SARS-CoV nucleocapsid protein. Genomics Proteomics Bioinformatics **1:** 198–206.
23. HUANG, Q., L. YU, A.M. PETROS, *et al.* 2004. Structure of the N-terminal RNA-binding domain of the SARS CoV nucleocapsid protein. Biochemistry **43:** 6059–6063.
24. HSIEH, P.K., S.C. CHANG, C.C. HUANG, *et al.* 2005. Assembly of severe acute respiratory syndrome coronavirus RNA packaging signal into virus-like particles is nucleocapsid dependent. J. Virol. **79:** 13848–13855.
25. SURJIT, M., B. LIU, P. KUMAR, *et al.* 2004. The nucleocapsid protein of the SARS coronavirus is capable of self-association through a C-terminal 209 amino acid interaction domain. Biochem. Biophys. Res. Commun. **317:** 1030–1036.
26. YU, I.M., C.L. GUSTAFSON, J. DIAO, *et al.* 2005. Recombinant severe acute respiratory syndrome (SARS) coronavirus nucleocapsid protein forms a dimer through its C-terminal domain. J. Biol. Chem. **280:** 23280–23286.
27. LUO, C., H. LUO, S. ZHENG, *et al.* 2004. Nucleocapsid protein of SARS coronavirus tightly binds to human cyclophilin A. Biochem. Biophys. Res. Commun. **321:** 557–565.
28. GAMBLE, T.R., F.F. VAJDOS, S. YOO, *et al.* 1996. Crystal structure of human cyclophilin A bound to the amino-terminal domain of HIV-1 capsid. Cell **87:** 1285–1294.

29. LUO, H., Q. CHEN, J. CHEN, *et al.* 2005. The nucleocapsid protein of SARS coronavirus has a high binding affinity to the human cellular heterogeneous nuclear ribonucleoprotein A1. FEBS Lett. **579:** 2623–2628.
30. XU, R.M., L. JOKHAN, X. CHENG, *et al.* 1997. Crystal structure of human UP1, the domain of hnRNP A1that contains two RNA-recognition motifs. Structure **5:** 559–570.
31. WANG, Y. & X. ZHANG. 1999. The nucleocapsid protein of coronavirus mouse hepatitis virus interacts with the cellular heterogeneous nuclear ribonucleoprotein A1 *in vitro* and *in vivo*. Virology **265:** 96–109.
32. HE, R., A. LEESON, A. ANDONOV, *et al.* 2003. Activation of AP-1 signal transduction pathway by SARS coronavirus nucleocapsid protein. Biochem. Biophys. Res. Commun. **311:** 870–876.
33. SURJIT, M., B. LIU, V.T. CHOW & S.K. LAL. 2006. The nucleocapsid protein of sars-coronavirus inhibits the activity of cyclin-CDK complex and blocks S phase progression in mammalian cells. J. Biol. Chem. **281:** 10669–10681.
34. SURJIT, M., B. LIU, S. JAMEEL, *et al.* 2004. The SARS coronavirus nucleocapsid protein induces actin reorganization and apoptosis in COS-1 cells in the absence of growth factors. Biochem. J. **383:** 1–6.
35. TIMANI, K.A., Q. LIAO, L. YE, *et al.* 2005. Nuclear/nucleolar localization properties of C-terminal nucleocapsid protein of SARS coronavirus. Virus Res. **114:** 23–34.
36. SURJIT, M., R. KUMAR, R.N. MISHRA, *et al.* 2005. The severe acute respiratory syndrome coronavirus nucleocapsid protein is phosphorylated and localizes in the cytoplasm by 14-3-3-mediated translocation. J. Virol. **79:** 11476–11486.
37. WOOTTON, S.K., R.R. ROWLAND & D. YOO. 2002. Phosphorylation of the porcine reproductive and respiratory syndrome virus nucleocapsid protein. J. Virol. **76:** 10569–10576.
38. ROWLAND, R.R., V. CHAUHAN, Y. FANG, *et al.* 2005. Intracellular localization of the severe acute respiratory syndrome coronavirus nucleocapsid protein: absence of nucleolar accumulation during infection and after expression as a recombinant protein in vero cells. J. Virol. **79:** 11507–11512.
39. FANG, X., L. YE, K.A. TIMANI, *et al.* 2005. Peptide domain involved in the interaction between membrane protein and nucleocapsid protein of SARS-associated coronavirus. J. Biochem. Mol. Biol. **38:** 381–385.
40. VOSS, D., A. KERN, E. TRAGGIAI, *et al.* 2006. Characterization of severe acute respiratory syndrome coronavirus membrane protein. FEBS Lett. **580:** 968–973.
41. NAL, B., C. CHAN, F. KIEN, *et al.* 2005. Differential maturation and subcellular localization of severe acute respiratory syndrome coronavirus surface proteins S, M and E. J. Gen. Virol. **86:** 1423–1434.
42. LIAO, Y., Q. YUAN, J. TORRES, *et al.* 2006. Biochemical and functional characterization of the membrane association and membrane permeabilizing activity of the severe acute respiratory syndrome coronavirus envelope protein. Virology **349:** 264–275.
43. LAW, P.T., C.H. WONG, T.C. AU, *et al.* 2005. The 3a protein of severe acute respiratory syndrome-associated coronavirus induces apoptosis in Vero E6 cells. J. Gen. Virol. **86:** 1921–1930.
44. TAN, Y.-J. 2005. The severe acute respiratory syndrome (SARS) coronavirus 3a protein may function as a modulator of the trafficking properties of the spike protein. Virol. J. **2:** 5.

45. TAN, Y.-J., P.-Y. THAM, D.Z.L. CHAN, *et al.* 2005. The severe acute respiratory syndrome coronavirus 3a protein up-regulates expression of fibrinogen in lung epithelial cells. J. Virol. **79:** 10083–10087.
46. LAW, P.T.W., C.-H. WONG, T.C.C. AU, *et al.* 2005. The 3a protein of severe acute respiratory syndrome-associated coronavirus induces apoptosis in Vero E6 cells. J. Gen. Virol. **86:** 1921–1930.
47. TAN, Y.-J., B.C. FIELDING, P.-Y. GOH, *et al.* 2004. Overexpression of 7a, a protein specifically encoded by severe acute respiratory syndrome coronavirus, induces apoptosis via a caspase-dependent pathway. J. Virol. **78:** 14043–14047.
48. LU, W., B.-J. ZHENG, K. XU, *et al.* 2006. Severe acute respiratory syndrome-associated coronavirus 3a protein forms an ion channel and modulates virus release. Proc. Natl. Acad. Sci. USA **103:** 12540–12545.
49. LUO, H., J. CHEN, K. CHEN, *et al.* 2006. Carboxyl terminus of severe acute respiratory coronavirus nucleocapsid protein: self-association analysis and nucleic acid binding characteristics. Biochemistry **45:** 11827–11835.
50. FANG, X., L.-B. YE, Y. ZHANG, *et al.* 2006. Nucleocapsid amino acids 211 to 254, in particular, tetrad glutamines, are essential for the interaction between the nucleocapsid and membrane proteins of SARS-associated coronavirus. J. Microbiol. **44:** 577–580.
51. ZUNIGA, S., I. SOLA, J.L. MORENO, *et al.* 2007. Coronavirus nucleocapsid protein is an RNA chaperone. Virology **357:** 215–227.

Molecular Characteristics of the Nipah Virus Glycoproteins

SANDRA DIEDERICH AND ANDREA MAISNER

Institute of Virology, Philipps University of Marburg, Germany

ABSTRACT: Nipah virus (NiV) is a highly pathogenic paramyxovirus, which emerged in 1998 from fruit bats in Malaysia and caused an outbreak of severe respiratory disease in pigs and fatal encephalitis in humans with high mortality rates. In contrast to most paramyxoviruses, NiV can infect a large variety of mammalian species. Due to this broad host range, its zoonotic potential, its high pathogenicity for humans, and the lack of effective vaccines or therapeutics, NiV was classified as a biosafety level 4 pathogen. This article provides an overview of the molecular characteristics of NiV focusing on the structure, functions, and unique biological properties of the two NiV surface glycoproteins, the receptor-binding G protein, and the fusion protein F. Since viral glycoproteins are major determinants for cell tropism and virus spread, a detailed knowledge of these proteins can help to understand the molecular basis of viral pathogenicity.

KEYWORDS: Nipah virus; viral glycoproteins; G protein; fusion protein; proteolytic cleavage

INTRODUCTION

Nipah virus (NiV), a highly pathogenic paramyxovirus, was isolated in 1999 after an outbreak of fatal encephalitis among pig farmers in Malaysia and Singapore.[1,2] Old World fruit bats of the genus *Pteropus* have been implicated as the natural host of NiV.[3,4] In 1997, NiV was transmitted from bats to pigs and then to humans suggesting that the pig was required as amplifying host. But in more recent outbreaks in Bangladesh (2001–2005), NiV was also transmitted directly from bats to humans and even human-to-human transmission was described.[5,6] Although pulmonary manifestations were observed in the infected individuals, most patients presented with a severe encephalitic syndrome. Blood vessels appear to be the key targets of injury in many organs, the central nervous system being the most severely affected in humans.[7,8] In addition to the acute infection, NiV can persist in the brain causing relapsed

Address for correspondence: Dr. Andrea Maisner, Philipps-Universität Marburg, Institut für Virologie, Hans-Meerwein-Str. 2, 35043 Marburg, Germany. Voice: 06421-28-65360; fax: 06421-28-65381. maisner@staff.uni-marburg.de

Ann. N.Y. Acad. Sci. 1102: 39–50 (2007).
doi: 10.1196/annals.1408.003

encephalitis after months or years after the initial infection.[9] Due to (1) its broad host and cell tropism, NiV infects several mammalians *in vivo* and replicates in a wide range of cultured cells,[10,11] (2) its high virulence in most animal species and humans (in recent outbreaks of human disease in Bangladesh the case fatality rate was 75%),[5] and (3) the lack of therapeutic treatment and a vaccine, NiV has been classified as biosafety 4 level agent.

NIV CLASSIFICATION

NiV belongs to the order *Mononegavirales* and is a member of the *Paramyxoviridae* family. Because of unique biological and genetic characteristics, NiV and the closely related Hendra virus (HeV) form a new genus *Henipavirus* within the subfamily *Paramyxovirinae* (FIG. 1).[12] The most striking unusual biological features of henipaviruses are their zoonotic potential and the broad host tropism including pigs, dogs, cats, horses, guinea pigs, hamsters, bats, and humans.[7,11,13,14] The most prominent unique molecular characteristic of the henipaviruses is their large genome. Whereas the other members of the *Paramyxoviridae* have genomes of approximately 15,500 nucleotides in length, the genome of henipaviruses is much larger (18,246 nucleotides in the case of NiV). This is mainly due to exceptional large untranslated regions at the 3′-end of most genes and a phosphoprotein that is more than 100 amino acids longer than the respective protein of other paramyxoviruses.[15]

GENOMIC ORGANIZATION AND VIRUS-ENCODED PROTEINS

NiV is an enveloped virus containing a single-stranded nonsegmented RNA genome of negative polarity. The genome encodes for six structural proteins in the gene order 3′-leader-N-P-M-F-G-L-trailer-5′ (FIG. 2). The viral envelope is derived from the host cell plasma membrane and surrounds the viral ribonucleocapsid (RNP), which consists of the viral genome tightly encapsidated by the viral nucleoprotein (N), the phosphoprotein (P), and the polymerase (L). As with many paramyxoviruses, the RNA is encapsidated by the N protein in a ratio of one N for every six nucleotides and the NiV RNA polymerase replicates only viral genomes counting 6n + 0 nucleotides. Because an exact

Order	Family	Subfamily	Genus	
MONONEGAVIRALES				
	Paramyxoviridae			
		Paramyxovirinae		
			Henipavirus	
				Nipah virus (NiV)
				Hendra virus (HeV)

FIGURE 1. Classification of NiV.

nucleotide-N match at the RNA 3′-end is required for the formation of an active replication promoter, the NiV genome must conform to the "rule of six."[16,17]

Protruding from the NiV lipid envelope are two surface glycoproteins, the receptor-binding G protein and the fusion (F) protein. The sixth structural protein is the matrix protein (M) that is located at the inner surface of the lipid bilayer and mediates the contact between the RNP and the surface glycoproteins (FIG. 2). In addition to the six structural proteins, NiV encodes for three additional proteins, which are transcribed from the P gene and are either generated by RNA editing (V and W protein) or are transcribed from an overlapping reading frame (C protein). In contrast to the P protein, the P gene products V, W, and C are not incorporated into the virus particle and likely act as inhibitors of the interferon signaling in NiV-infected cells.[18–21]

OVERVIEW ON THE NIV REPLICATION CYCLE

Like all paramyxoviruses, NiV replicates in the cytoplasm of the infected cell. After binding of the virion to the cell surface via the G protein, fusion of the virus envelope and the plasma membrane takes place. Thereby, the viral RNP is liberated into the cytoplasm and the viral replication cycle starts with the transcription of the genomic RNA into mRNAs by the viral polymerase L. Since the polymerase detaches at each intergenic junction and reinitiates

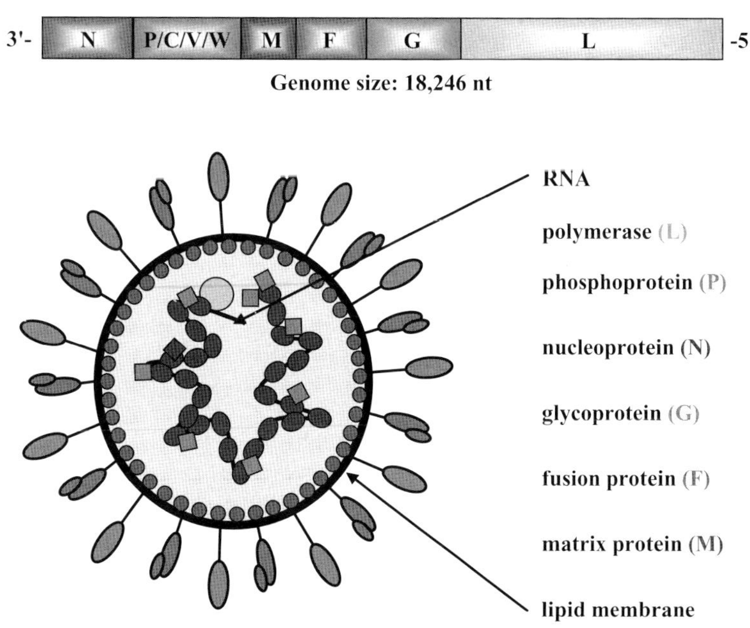

FIGURE 2. Genome organization and structural components of NiV.

at the 3′-end of the next gene a transcription gradient from the N to the L gene is generated. Thus, the amount of mRNAs and that of the corresponding viral proteins depends on the distance of the gene from the 3′ -promotor. The mRNAs for the two surface glycoproteins G and F are translated at ribosomes associated with the endoplasmic reticulum (ER). The glycoproteins are co-translationally inserted into the ER membrane and then transported through the secretory pathway to the plasma membrane. All other NiV proteins (M, N, P, and L) are synthesized at ribosomes in the cytoplasm. As soon as enough N protein is produced, the polymerase switches from transcription (mRNA synthesis) to replication (synthesis of full-length antigenomes and genomes). The newly generated RNA genomes are immediately encapsidated by the N proteins and RNPs are formed. The final assembly is believed to occur at the plasma membrane where subsequent budding of the new virions takes place.[17]

NIV SURFACE GLYCOPROTEINS

The two surface glycoproteins G and F of NiV are essentially required for virus entry into host cells and are therefore the major determinants for the broad species and host cell tropism of NiV *in vivo* and *in vitro*. The G protein mediates attachment to the cellular receptor and the fusion protein F is responsible for virus entry into host cells by initiating pH-independent fusion of the viral and cellular membranes.[17] If coexpressed with the G protein on the surface of infected cells or on cells expressing plasmid-encoded NiV glycoproteins, the F protein mediates fusion with adjacent cells, resulting in the formation of multinucleated syncytia.[10,22] The glycoprotein-mediated cell-to-cell fusion is an important means for the virus to spread directly from infected to uninfected neighboring cells in the absence of virus budding and significantly influences the cytopathogenicity of an NiV infection.[23]

Attachment Protein G

The G protein is a type II membrane glycoprotein of 602 amino acids and functions primarily as receptor-binding protein.[15] Besides its function in receptor binding, the G protein is essentially required for F-mediated fusion of lipid membranes.[10] In contrast to most other paramyxoviruses, the NiV G protein lacks hemagglutinating and neuraminidase activities and does not bind to carbohydrate moieties.[24,25] Though the idea that NiV uses a cellular surface protein for virus entry had already been proposed some years ago,[10,26] this has been proven only recently. Two independent groups identified ephrin B2 as cellular receptor for henipaviruses.[27,28] Ephrin B2 is a cell-surface glycoprotein, which normally functions as ligand for Eph receptors and plays a key role during development especially in the vascular and nervous system.[29–32] Ephrin B2 is widely expressed in different cells and tissues of many species but most abundantly in smooth muscle cells, neurons, and endothelial cells.[33–35] Furthermore, ephrin B2 is upregulated in certain tumor cells.[36] Thus, the receptor distribution provides a good explanation for the cell and tissue tropism of NiV *in vivo* as well as for the broad cell tropism in cultured tumor cells. Very recently, ephrin B3 has been proposed as alternative receptor for NiV. As ephrin B3 is expressed in some regions of the brain where the main entry receptor ephrin B2 is not present, it may account for some of the pathology in the central nervous system seen in NiV patients.[37]

Fusion Protein F

Structure of the F Protein

Like the F proteins of other *Paramyxoviridae*,[17] the NiV F protein is predicted to be a type I transmembrane protein. It consists of 546 amino acids with a molecular mass of 60 kDa.[15] After synthesis and cotranslational glycosylation in the ER, the F_0 precursor protein oligomerizes into trimers.[25] The F trimers are then transported through the Golgi apparatus to the plasma membrane. Before incorporation into new virions, the F_0 precursor is cleaved into the two disulfide-linked subunits F_1 and F_2. The F_1 cleavage product derived from the C terminus of F_0 contains several functional domains: The cytoplasmic tail located at the very C terminus has a length of 28 amino acids and has recently been shown to contain a tyrosine-based amino acid motif, which mediates constitutive endocytosis of the NiV F protein.[38] The cytoplasmic domain is followed by the transmembrane domain, which is responsible for anchoring the F protein in lipid membranes. The N terminus of the F_1 subunit is formed by a stretch of about 20 hydrophobic amino acids, the so-called fusion peptide. The fusion peptide is highly conserved among all paramyxoviruses and is indispensable for the biological activity of the F protein. Its role is to insert into the target membrane thereby initiating the fusion process.[39,40] Interestingly, the fusion peptide at the N terminus of the F_1 subunit starts with a leucine, whereas all other paramyxoviruses with the exception of some avirulent strains of Newcastle disease virus have a phenylalanine (TABLE 1).[15] The biological consequences of this conserved amino acid exchange are still unclear.

As reported for other paramyxoviral F proteins, the NiV F_1 subunit contains two α-helical domains referred to as heptad repeats. Heptad repeat-1 is

TABLE 1. Proteolytic activation and spread of paramyxoviruses

Virus	Sequence upstream of the cleavage site	Fusion peptide	Activating proteases	Virus spread *in vivo*
HPIV-1	D-N-P-Q-T-**R**	F-F-G-A-V	TLPs	Restricted
Sendai virus	D-V-P-Q-S-**R**	F-F-G-A-V	TLPs	Restricted
Mumps virus	S-**R**-**R**-H-**K**-**R**	F-A-G-I-A	Furin	Systemic
SV5	T-**R**-**R**-**R**-**R**-**R**	F-A-G-V-V	Furin	Systemic
Virulent NDV	G-**R**-**R**-Q-**K**-**R**	F-I-G-A-I	Furin	Systemic
Avirulent NDV	G-G-**R**-Q-G-**R**	L-I-G-A-I	TLPs	Restricted
Measles virus	S-**R**-**R**-H-**K**-**R**	F-A-G-V-V	Furin	Systemic
CDV	G-**R**-**R**-Q-**R**-**R**	F-A-G-V-V	Furin	Systemic
RPV	S-**R**-**R**-H-**K**-**R**	F-A-G-V-V	Furin	Systemic
Hendra virus	L-V-G-D-V-**K**	L-A-G-V-V	Cathepsin L	Systemic
Nipah virus	L-V-G-D-V-**R**	L-A-G-V-I	Cathepsin L	Systemic

Adapted from Klenk and Garten[60]; Harcourt *et al.*,[15].
HPIV-1 =Human Parainfluenza virus 1; SV5 = Simian virus 5; NDV = Newcastle disease virus; CDV = Canine distemper virus; RPV = Rinderpest virus; TLPs = Trypsin-like proteases.

located proximal to the fusion peptide and heptad repeat-2 precedes the transmembrane domain.[26,41] Upon fusion with cellular membranes, the two heptad repeats within paramyxoviral F proteins are known to form a tight complex, in which the N-terminal heptad repeat forms a trimeric-coiled coil, with the C-terminal heptad repeat forming helical regions that buttress the coiled-coil in an antiparallel manner. This structure is also called a 6-helix bundle.[42,43] It has also been shown for the NiV F protein that the heptad repeat domains interact with each other and form the typical 6-helix coiled-coil bundles during or immediately following fusion.[41] Peptides corresponding to the C-terminal heptad repeat of NiV F are potent inhibitors of NiV-mediated membrane fusion in a recombinant expression system and in NiV-infected cells.[44]

Glycosylation of the F Protein

N-glycosylation of fusion proteins might be of importance for the structural and functional integrity. For many viral glycoproteins, it has been shown that N-glycans are important structural components that affect folding and transport[45–48] as well as the activity,[49–52] and immunological properties of these molecules.[53–55] The NiV F protein as well as the closely related HeV F protein contains 5 potential N-glycosylation consensus sites (N-X-S/T, in which X can be any amino acid except proline), 3 located in the F_2 subunit, and 2 of them in the F_1 subunit (FIG. 3).[15,56] Analysis of the glycosylation pattern revealed that 2 of the 3 sites in the F_2 subunit (N67 and N99) and both sites in the F_1 subunit (N414 and N464) are used for N-linked glycosylation in both

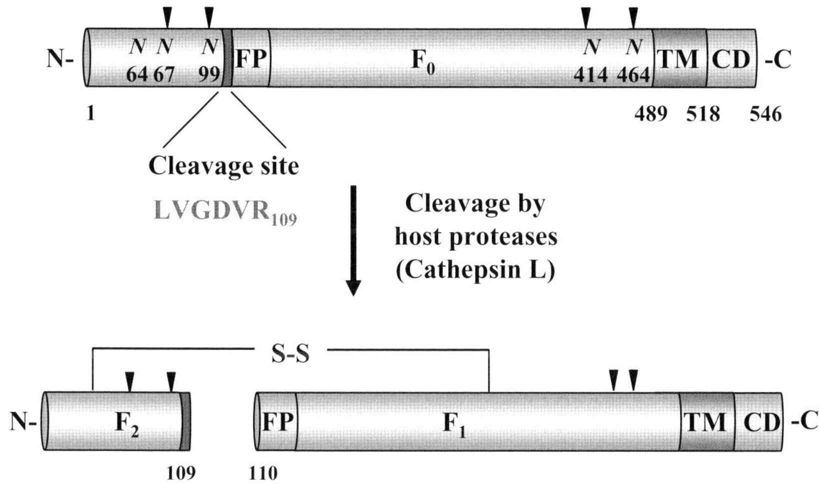

F_0	Uncleaved precursor F protein	**TM**	Transmembrane domain
F_1	C-terminal F cleavage product	**CD**	Cytoplasmic domain
F_2	N-terminal F cleavage product	*N*	Potential glycosylation site
FP	Fusion peptide	▼	N-glycan

FIGURE 3. Schematic representation of the NiV fusion protein.

henipavirus F proteins.[57,58] Three of the four N-glycans have been shown to be complex glycosylated (oligosaccharides linked to N67, N99, and N414), whereas the glycan attached to N464 is of the high-mannose type.[57] Only the N-linked modification at N414 is absolutely critical for protein folding and transport to the cell surface, whereas absence of the oligosaccharide attached to N464, which lies within heptad repeat-2, does not impair fusion activity even though cell-surface expression is slightly decreased.[57,58] This is in contrast to many other paramyxoviral fusion proteins containing a carbohydrate within the membrane-proximal heptad repeat. Here, the glycan is essentially required for fusion activity,[45,49,52] suggesting a different fusion mechanism promoted by henipavirus F proteins. Interestingly, the N-glycan at N99, which is located only 10 amino acids upstream of the cleavage site (arginine at position 109), has no implications on proteolytic F-activation. Such proximity of a N-glycan to the cleavage site is so far only known for the F protein of bovine respiratory syncytial virus.[52] Taken together, the individual N-linked oligosaccharides of the NiV F protein clearly differ in terms of their importance for protein folding, intracellular transport, and fusion activity. Very recently it was described that N-glycosylation of the F protein influences the association of the F and G protein and thus the fusogenicity of the NiV glycoprotein complex. Furthermore, the N-glycans were supposed to protect NiV from antibody neutralization but their exact role in modulating the antigenicity of the molecule has still to be elucidated.[59]

Exceptional Proteolytic Activation of the NiV F Protein

It is known that proteolytic cleavage of the fusion protein of paramyxoviruses at basic amino acid residues (arginine or lysine) is a prerequisite for virus infectivity. In addition to the availability of viral receptors, differences in the distribution of F-protein-activating proteases are crucial factors in determining host range, tissue tropism, and pathogenicity.[60–63] In analogy to other F proteins, the inactive NiV precursor protein F_0 has to be activated by cleavage into F_1 and F_2 thereby releasing the fusion peptide at the N terminus of F_1 (FIG. 3). Only F proteins, which are correctly cleaved directly upstream of the fusion peptide, can promote fusion of the viral envelope and cellular membranes during virus entry or fusion of neighboring cells at later stages of infection. So far, two mechanisms of F-activation are known: Paramyxoviruses that cause systemic infections *in vivo*, for example, measles or mumps virus, usually possess F proteins with multiple basic amino acid residues at the cleavage site (TABLE 1). These F proteins can be activated ubiquitously during transport through the secretory pathway by the Golgi protease furin. Because of the intracellular cleavage event, these F proteins reach the cell surface as fusion-active molecules. In contrast, some F proteins, for example, Sendai virus or human parainfluenza virus-1 F protein, contain only one basic amino acid residue at the cleavage site and usually cannot be activated intracellularly but have to be cleaved by extracellular trypsin-like proteases after arrival at the cell surface. Consequently, growth of these viruses *in vitro* depends on the addition of exogenous trypsin to the cell culture medium and moreover, replication of these viruses *in vivo* is restricted to the respiratory tract where suitable proteases like tryptase Clara or miniplasmin are known to be expressed.[64,65] Considering the fact that NiV causes a systemic infection *in vivo* and replicates *in vitro* in the absence of exogenous trypsin, it is surprising that the F protein has no multibasic cleavage site but only one single basic arginine upstream of the fusion peptide.[10,22,23] In the closely related HeV F protein, the arginine at the cleavage site is replaced by lysine (TABLE 1). Replication of NiV in cell culture even in the absence of serum proteases showed that F cleavage does not depend on any ubiquitous exogenous protease but on a cellular protease.[66] The surprising finding that ubiquitous activation of the NiV F proteins was not abolished by replacing the basic arginine residue at the cleavage site by a nonbasic amino acid indicates that the activating protease is no member of the trypsin- or furin-like protease families known to activate other paramyxoviral F proteins.[23] Inhibitor studies revealed that not only cleavage of the NiV F protein is exceptional with respect to protease usage but also that the subcellular localization where proteolytic activation takes place is strikingly different. Blocking of protein transport through the secretory pathway or selective inhibition of endocytosis as well as changes in the intracellular pH drastically reduced proteolytic cleavage of the NiV F protein. This demonstrates that activation requires F transport to the cell surface and subsequent clathrin-mediated endocytosis followed by proteolytic processing in the acidic milieu of the early or recycling endosomal compartment.[66] To ensure constitutive internalization and thus cleavage, the cytoplasmic tail of the NiV F protein contains a classical endocytosis signal of the YXX-type.[38] The observation that in the absence of a functional internalization signal F cleavage was drastically downregulated confirmed that constitutive endocytosis is of critical importance for the biological activation of the fusion protein.[66] In full agreement to what has been previously described for the NiV F protein, the closely related HeV F protein has been shown to also depend on exceptional sequence requirements at the cleavage site as well as to require endocytosis to get activated.[67–69] The recent reports demonstrating that the endosomal protease cathepsin L is involved in cleavage of both the HeV and NiV F protein[70,71] further support the conclusion that proteolytic processing of henipavirus F proteins depends on an activation mechanism completely different from what is known for other paramyxoviruses.

REFERENCES

1. CHUA, K.B. 2003. Nipah virus outbreak in Malaysia. J. Clin. Virol. **26:** 265–275.
2. CHUA, K.B. *et al.* 2000. Nipah virus: a recently emergent deadly paramyxovirus. Science **288:** 1432–1435.
3. YOB, J.M. *et al.* 2001. Nipah virus infection in bats (order Chiroptera) in peninsular Malaysia. Emerg. Infect. Dis. **7:** 439–441.
4. CHUA, K.B. *et al.* 2002. Isolation of Nipah virus from Malaysian island flying-foxes. Microbes Infect. **4:** 145–151.
5. ANONYMOUS. 2004. Person-to-person transmission of Nipah virus during outbreak in Faridpur district, 2004. Health Sci. Bull. **2:** 5–9.
6. CHOI, C. 2004. Nipah's return. The lethal "flying fox" virus may spread between people. Sci. Am. **291:** 21A, 22.
7. CHUA, K.B. *et al.* 1999. Fatal encephalitis due to Nipah virus among pig-farmers in Malaysia. Lancet **354:** 1257–1259.
8. WONG, K.T. *et al.* 2002. Nipah virus infection: pathology and pathogenesis of an emerging paramyxoviral zoonosis. Am. J. Pathol. **161:** 2153–2167.
9. TAN, C.T. *et al.* 2002. Relapsed and late-onset Nipah encephalitis. Ann. Neurol. **51:** 703–708.
10. BOSSART, K.N. *et al.* 2002. Membrane fusion tropism and heterotypic functional activities of the Nipah virus and Hendra virus envelope glycoproteins. J. Virol. **76:** 11186–11198.
11. HOOPER, P. *et al.* 2001. Comparative pathology of the diseases caused by Hendra and Nipah viruses. Microbes Infect. **3:** 315–322.

12. WANG, L.F. *et al.* 2000. The exceptionally large genome of Hendra virus: support for creation of a new genus within the family Paramyxoviridae. J Virol. **74:** 9972–9979.
13. FIELD, H. *et al.* 2001. The natural history of Hendra and Nipah viruses. Microbes Infect. **3:** 307–314.
14. WONG, K.T. *et al.* 2003. A golden hamster model for human acute Nipah virus infection. Am. J. Pathol. **163:** 2127–2137.
15. HARCOURT, B.H. *et al.* 2000. Molecular characterization of Nipah virus, a newly emergent paramyxovirus. Virology **271:** 334–349.
16. HALPIN, K. *et al.* 2004. Nipah virus conforms to the rule of six in a minigenome replication assay. J. Gen. Virol. **85:** 701–707.
17. LAMB, R.A. & D. KOLAKOVSKY. 2001. Paramyxoviridae: the viruses and their replication. *In* Fields Virology. Fourth edition. D.M. Knipe *et al.*, Eds.: 1305–1340. Lippincott, Williams and Williams. Washington, DC.
18. RODRIGUEZ, J.J., J.P. PARISIEN & C.M. HORVATH. 2002. Nipah virus V protein evades alpha and gamma interferons by preventing STAT1 and STAT2 activation and nuclear accumulation. J. Virol. **76:** 11476–11483.
19. RODRIGUEZ, J.J., C.D. CRUZ & C.M. HORVATH. 2004. Identification of the nuclear export signal and STAT-binding domains of the Nipah virus V protein reveals mechanisms underlying interferon evasion. J. Virol. **78:** 5358–5367.
20. SHAW, M.L. *et al.* 2004. Nipah virus V and W proteins have a common STAT1-binding domain yet inhibit STAT1 activation from the cytoplasmic and nuclear compartments, respectively. J. Virol. **78:** 5633–5641.
21. SHAW, M.L. *et al.* 2005. Nuclear localization of the Nipah virus W protein allows for inhibition of both virus- and toll-like receptor 3-triggered signaling pathways. J Virol. **79:** 6078–6088.
22. TAMIN, A. *et al.* 2002. Functional properties of the fusion and attachment glycoproteins of Nipah virus. Virology **296:** 190–200.
23. MOLL, M. *et al.* 2004. Ubiquitous activation of the Nipah virus fusion protein does not require a basic amino acid at the cleavage site. J. Virol. **78:** 9705–9712.
24. MICHALSKI, W.P. *et al.* 2000. The cleavage activation and sites of glycosylation in the fusion protein of Hendra virus. Virus Res. **69:** 83–93.
25. EATON, B.T. *et al.* 2004. Henipaviruses: recent observations on regulation of transcription and the nature of the cell receptor. Arch. Virol. Suppl. 122–131.
26. BOSSART, K.N. *et al.* 2001. Functional expression and membrane fusion tropism of the envelope glycoproteins of Hendra virus. Virology **290:** 121–135.
27. BONAPARTE, M.I. *et al.* 2005. Ephrin-B2 ligand is a functional receptor for Hendra virus and Nipah virus. Proc. Natl. Acad. Sci. USA **102:** 10652–10657.
28. NEGRETE, O.A. *et al.* 2005. EphrinB2 is the entry receptor for Nipah virus, an emergent deadly paramyxovirus. Nature **436:** 401–405.
29. DAVIS, S. *et al.* 1994. Ligands for EPH-related receptor tyrosine kinases that require membrane attachment or clustering for activity. Science **266:** 816–819.
30. FLANAGAN, J.G. & P. VANDERHAEGHEN. 1998. The ephrins and Eph receptors in neural development. Annu. Rev. Neurosci. **21:** 309–345.
31. FRISEN, J., J. HOLMBERG & M. BARBACID. 1999. Ephrins and their Eph receptors: multitalented directors of embryonic development. EMBO J. **18:** 5159–5165.
32. GALE, N.W. & G.D. YANCOPOULOS. 1997. Ephrins and their receptors: a repulsive topic? Cell Tissue Res. **290:** 227–241.
33. GALE, N.W. *et al.* 2001. Ephrin-B2 selectively marks arterial vessels and neovascularization sites in the adult, with expression in both endothelial and smooth-muscle cells. Dev. Biol. **230:** 151–160.

34. SHIN, D. *et al.* 2001. Expression of ephrinB2 identifies a stable genetic difference between arterial and venous vascular smooth muscle as well as endothelial cells, and marks subsets of microvessels at sites of adult neovascularization. Dev. Biol. **230:** 139–150.
35. WANG, H.U., Z.F. CHEN & D.J. ANDERSON. 1998. Molecular distinction and angiogenic interaction between embryonic arteries and veins revealed by ephrin-B2 and its receptor Eph-B4. Cell **93:** 741–753.
36. HAFNER, C. *et al.* 2004. Differential gene expression of Eph receptors and ephrins in benign human tissues and cancers. Clin. Chem. **50:** 490–499.
37. NEGRETE, O.A. *et al.* 2006. Two key residues in ephrinB3 are critical for its use as an alternative receptor for Nipah virus. PLoS Pathog. **2:** 78–86.
38. VOGT, C. *et al.* 2005. Endocytosis of the Nipah virus glycoproteins. J. Virol. **79:** 3865–3872.
39. GETHING, M.J., J.M. WHITE & M.D. WATERFIELD. 1978. Purification of the fusion protein of Sendai virus: analysis of the NH2-terminal sequence generated during precursor activation. Proc. Natl. Acad. Sci. USA **75:** 2737–2740.
40. EPAND, R.M. 2003. Fusion peptides and the mechanism of viral fusion. Biochim. Biophys. Acta **1614:** 116–121.
41. XU, Y. *et al.* 2004. Basis for fusion inhibition by peptides: analysis of the heptad repeat regions of the fusion proteins from Nipah and Hendra viruses, newly emergent zoonotic paramyxoviruses. Biochem. Biophys. Res. Commun. **315:** 664–670.
42. DUTCH, R.E., T.S. JARDETZKY & R.A. LAMB. 2000. Virus membrane fusion proteins: biological machines that undergo a metamorphosis. Biosci. Rep. **20:** 597–612.
43. EARP, L.J. *et al.* 2005. The many mechanisms of viral membrane fusion proteins. Curr. Top. Microbiol. Immunol. **285:** 25–66.
44. BOSSART, K.N. *et al.* 2005. Inhibition of Henipavirus fusion and infection by heptad-derived peptides of the Nipah virus fusion glycoprotein. Virol J. **2:** 57.
45. BAGAI, S. & R.A. LAMB. 1995. Individual roles of N-linked oligosaccharide chains in intracellular transport of the paramyxovirus SV5 fusion protein. Virology **209:** 250–256.
46. DOMS, R.W. *et al.* 1993. Folding and assembly of viral membrane proteins. Virology **193:** 545–562.
47. ROBERTS, P.C., W. GARTEN & H.D. KLENK. 1993. Role of conserved glycosylation sites in maturation and transport of influenza A virus hemagglutinin. J. Virol. **67:** 3048–3060.
48. TAMURA, T. *et al.* 2002. N-linked oligosaccharide chains of Sendai virus fusion protein determine the interaction with endoplasmic reticulum molecular chaperones. FEBS Lett. **513:** 153–158.
49. MCGINNES, L. *et al.* 2001. Carbohydrate modifications of the NDV fusion protein heptad repeat domains influence maturation and fusion activity. Virology **283:** 332–342.
50. SEGAWA, H. *et al.* 2000. Functional analysis of the individual oligosaccharide chains of sendai virus fusion protein. J. Biochem. (Tokyo) **128:** 65–72.
51. WAGNER, R. *et al.* 2000. Interdependence of hemagglutinin glycosylation and neuraminidase as regulators of influenza virus growth: a study by reverse genetics. J. Virol. **74:** 6316–6323.
52. ZIMMER, G., I. TROTZ & G. HERRLER. 2001. N-glycans of F protein differentially affect fusion activity of human respiratory syncytial virus. J. Virol. **75:** 4744–4751.

53. GARCIA-BEATO, R. *et al.* 1996. Host cell effect upon glycosylation and antigenicity of human respiratory syncytial virus G glycoprotein. Virology **221:** 301–309.
54. KOCH, M. *et al.* 2003. Structure-based, targeted deglycosylation of HIV-1 gp120 and effects on neutralization sensitivity and antibody recognition. Virology **313:** 387–400.
55. MUNK, K. *et al.* 1992. Carbohydrate masking of an antigenic epitope of influenza virus haemagglutinin independent of oligosaccharide size. Glycobiology **2:** 233–240.
56. WANG, L. *et al.* 2001. Molecular biology of Hendra and Nipah viruses. Microbes Infect. **3:** 279–287.
57. MOLL, M., A. KAUFMANN & A. MAISNER. 2004. Influence of N-glycans on processing and biological activity of the Nipah virus fusion protein. J. Virol. **78:** 7274–7278.
58. CARTER, J.R. *et al.* 2005. Role of N-linked glycosylation of the Hendra virus fusion protein. J. Virol. **79:** 7922–7925.
59. AGUILAR, H.C. *et al.* 2006. N-glycans on Nipah virus fusion protein protect against neutralization but reduce membrane fusion and viral entry. J. Virol. **80:** 4878–4889.
60. KLENK, H.D. & W. GARTEN. 1994. Activation cleavage of viral spike proteins by host proteases. *In* Cellular Receptors for Animal Viruses: 241–280. Cold Spring Harbor Laboratory. Cold Spring Harbor. New York.
61. MAISNER, A. *et al.* 2000. Recombinant measles virus requiring an exogenous protease for activation of infectivity. J. Gen. Virol. **81:** 441–449.
62. TASHIRO, M. *et al.* 1988. Characterization of a pantropic variant of Sendai virus derived from a host range mutant. Virology **165:** 577–583.
63. TASHIRO, M. *et al.* 1990. Organ tropism of Sendai virus in mice: proteolytic activation of the fusion glycoprotein in mouse organs and budding site at the bronchial epithelium. J. Virol. **64:** 3627–3634.
64. MURAKAMI, M. *et al.* 2001. Mini-plasmin found in the epithelial cells of bronchioles triggers infection by broad-spectrum influenza A viruses and Sendai virus. Eur. J. Biochem. **268:** 2847–2855.
65. TASHIRO, M. *et al.* 1992. Tryptase Clara, an activating protease for Sendai virus in rat lungs, is involved in pneumopathogenicity. J. Virol. **66:** 7211–7216.
66. DIEDERICH, S. *et al.* 2005. The Nipah virus fusion protein is cleaved within the endosomal compartment. J. Biol. Chem. **280:** 29899–29903.
67. PAGER, C.T., M.A. WURTH & R.E. DUTCH. 2004. Subcellular localization and calcium and pH requirements for proteolytic processing of the Hendra virus fusion protein. J. Virol. **78:** 9154–9163.
68. CRAFT, W.W., JR. & R.E. DUTCH. 2005. Sequence motif upstream of the Hendra virus fusion protein cleavage site is not sufficient to promote efficient proteolytic processing. Virology. **341:** 130–140.
69. MEULENDYKE, K.A. *et al.* 2005. Endocytosis plays a critical role in proteolytic processing of the Hendra virus fusion protein. J. Virol. **79:** 12643–12649.
70. PAGER, C.T. & R.E. DUTCH. 2005. Cathepsin L is involved in proteolytic processing of the Hendra virus fusion protein. J. Virol. **79:** 12714–12720.
71. PAGER, C.T. *et al.* 2006. A mature and fusogenic form of the Nipah virus fusion protein requires proteolytic processing by cathepsin L. Virology **346:** 251–257.

Envelope–Receptor Interactions in Nipah Virus Pathobiology

BENHUR LEE

Department of Microbiology, Immunology and Molecular Genetics, David Geffen School of Medicine at UCLA, Los Angeles, California, USA

ABSTRACT: Nipah (NiV) and Hendra (HeV) viruses are members of the newly defined *Henipavirus* genus of the *Paramyxoviridae.* Nipah virus (NiV) is an emergent paramyxovirus that causes fatal encephalitis in up to 70% of infected patients, and there is increasing evidence of human-to-human transmission. NiV is designated a priority pathogen in the NIAID Biodefense Research Agenda, and could be a devastating agent of *agro*bioterrorism if used against the pig farming industry. Endothelial syncytium is a pathognomonic feature of NiV infections, and is mediated by the fusion (F) and attachment (G) envelope glycoproteins. This review summarizes what is known about the pathophysiology of NiV infections, and documents the identification of the NiV receptor. EphrinB2, the NiV and HeV receptor, is expressed on endothelial cells and neurons, consistent with the known cellular tropism for NiV. We discuss how the identification of the henipahvirus receptor sheds light on the pathobiology of NiV infection, and how it will spur the rational development of effective therapeutics. In addition, ephrinB3, a related protein, can serve as an alternative receptor, and we suggest that differential usage of ephrinB2 versus B3 may explain the variant pathogenic profiles observed between NiV and HeV. Thus, identifying the NiV receptors opens the door for a more comprehensive analysis of the envelope–receptor interactions in NiV pathobiology. Finally, we also describe how galectin-1 (an innate immune defense lectin) can interact with specific N-glycans on the Nipah envelope fusion protein, underscoring the potential role that innate immune defense mechanisms may play against emerging pathogens.

KEYWORDS: paramyxovirus; ephrinb2; receptor; envelope; tropism

NIPAH VIRUS: A LETHAL EMERGING VIRAL PATHOGEN

In 1999, Nipah Virus (NiV) was discovered to be the causal agent of an outbreak of respiratory illness in pigs and encephalitis in humans.[1–3] In 2001–2005, NiV outbreaks occurred in Bangladesh,[4–7] where pigs were not found to

Address for correspondence: Dr. Benhur Lee, UCLA/MIMG, 3825 Mol. Sci. Bldg. 609 Charles E. Young Drive, East Los Angeles, CA 90095-1489. Voice: 310-206-8792; fax: 310-267-2580. bleebhl@ucla.edu

Ann. N.Y. Acad. Sci. 1102: 51–65 (2007).
doi: 10.1196/annals.1408.004

be infected, suggesting that the virus may have jumped directly from its natural host, the fruit bat or flying fox (genus *Pteropus*), to humans.[8–10] NiV infection typically leads to death from fatal encephalitis in humans, with a 40% to 70% mortality rate in the Malaysian and Bangladesh outbreaks, respectively.[11,12] The latter mortality rate is similar to that seen with viral agents, such as Ebola virus, and far exceeds the mortality associated with any of the recent emerging viral pathogens in South East Asia (e.g., SARS coronavirus and H5N1 influenza). While most clinically apparent cases in Malaysia were found in patients with direct exposure to infected pigs,[13] seroconversions and subclinical brain lesions were documented in exposed health care workers, demonstrating that human-to-human transmission of the virus is possible.[14] Indeed, there is stronger evidence of human-to-human transmission in the recent Bangladesh outbreaks.[11,15] NiV is a BSL-4 pathogen and is designated a Category C priority pathogen in the NIAID Biodefense Research Agenda.

BIOTERRORISM CONCERNS

NiV has been cited as a potential agent of bioterrorism due to several properties .[16] Its extreme pathogenicity (~40–70% mortality) exceeds that of smallpox (~30% mortality) and approaches that for Ebola (~40–90%). In addition to causing acute disease, 3–7% of infected patients exhibit a late onset or relapsed encephalitis months to years after the initial infection,[14,17,18] thereby increasing the risk of community exposure. This is particularly worrisome given that there is increasing evidence of human-to-human transmissions in the most recent outbreaks.[11] NiV can be grown to high titers *in vitro* ($>10^8$ IU/mL without concentration) ,[14] and thus, it has been suggested it can be weaponized as an aerosol.[16] In addition, NiV may be a more devastating agent of *economic* or *agrobioterrorism* when used against the pig farming industry. During the NiV outbreak in Malaysia, which affected 265 individuals, more than 1 million pigs were culled, resulting in economic losses that totaled far more than their export value of US$100 million.[16] In the United States, the total annual production value of hogs and pigs in 2002 exceeded $8 *billion,* with farms in just three states (Iowa, Minnesota, and North Carolina) accounting for 50% of the value .[19] If Nipah-like agents were to be intentionally released in any one of these three states, the economic devastation from the loss of production alone could easily be more than $1 billion. This is not an unrealistic scenario; while the mortality rate in pigs (~5%) is lower than in humans, the morbidity and transmission rate approaches 100%.[2,20] The observation that seroconversions and MRI-detectable brain lesions have been documented in exposed health care workers,[21] and in military personnel involved in the Malaysian culling operation, and that 30% of seroconverters went on to develop encephalitic sequelae,[22] further underscores the need to understand the pathogenesis of this disease.

NIPAH VIRUS BIOLOGY

As a member of the *Paramyxoviridae* family, the NiV contains a negative strand, nonsegmented, RNA genome, with a gene arrangement similar to other family members.[23] However, phlyogenetic analysis of NiV multiple genes suggests a clustering with the Hendra virus (HeV),[24,25] another paramyxovirus discovered a few years prior in Australia that caused an outbreak of severe respiratory syndrome in thoroughbred horses.[1,26] Functional and immunologic data also support the grouping of NiV and HeV into a new genus, as evidenced by the heterotypic complementation between their fusion (F) and attachment (G) envelope glycoproteins,[8,27] and the cross-reactivity between their respective sera.[24]

The NiV (and HeV) virus exhibits an unusually broad host range including humans, pigs, dogs, cats, horses, guinea pigs, hamsters, and fruit bats (its presumptive natural host).[2,28–30] Such broad host tropism is rare among extant paramyxoviruses. With the possible exception of fruit bats, disease mortality of all other hosts has been shown during natural or experimental infection.[30] However, despite evidence implicating pigs as the amplifying intermediate reservoir, the mortality in pigs (< 5%) is far lower than that in humans (40–70%),[1,2] suggesting that zoonotic transmission to humans has increased the pathogenicity of the virus. Understanding the determinants of such broad tropism, but narrower pathogenicity, is critical for understanding the etiology of the disease. Since cellular tropism is determined in large part by the specificity of receptor–envelope interactions, identifying the NiV and HeV receptor is the first step toward this goal. Indeed, identification of the cognate receptor for NiV and HeV entry will likely speed the rational development of effective therapeutics against this deadly emerging pathogen.

As the requirement for studying this virus under BSL-4 conditions would understandably limit our ability to gain insights into the biology of this new genus of *Paramyxoviridae*, researchers have taken advantage of the fact that endothelial cell syncytia formation is a pathogenic hallmark of Nipah viral disease (see below), and is strictly mediated by the F and G envelope glycoproteins.[8,21,27] Thus, we and others have generated systems and reagents to study the determinants involved in viral envelope-mediated entry at less than BSL-4 conditions.[8,27,31–33] Notably, fusion assays and pseudotyping the NiV and HeV envelope glycoproteins on a replication incompetent vesicular stomatitis virus (VSV) reporter virus, allow studies of envelope-mediated membrane fusion and viral entry at less than BSL-4 conditions.

NIPAH VIRUS PATHOPHYSIOLOGY

As mentioned above, NiV emerged among agricultural workers in Malaysia in 1998. Among the 265 reported cases in the first outbreak, there were 105 patient deaths.[18] While most patients presented with a severe acute encephalitic syndrome, many also had significant pulmonary symptoms. Viral particles were obtained from cerebral spinal fluid (CSF) and characterized morphologically, serologically, and genetically, and the new virus was named after the Nipah River Village where the first isolates were obtained.[2,3]

The careful pathological investigations of 32 patients who died of NiV infection, provide a reasonably comprehensive description of NiV pathophysiology.[18] Several critical points will be summarized here. First, NiV kills infected patients very rapidly; the average time from fever onset to death was 9.5 days, and only 4/32 patients survived >14 days before death. Patients presented with fever (100%) and other neurologic symptoms (drowsiness, headache, disorientation). About 40% patients presented with respiratory symptoms.

Second, a major cellular target of the NiV appears to be endothelial cells that line blood vessels. At autopsy, microscopic evaluation revealed widespread vasculitis, endothelial cell destruction, and focal perivascular necrosis in small vessels in the lung (62%), heart (31%), kidney (24%), and central nervous system (CNS; 80%), with the most severe damage seen to vessels in the CNS. Importantly, syncytial or multinucleated giant endothelial cells were seen in blood vessels in many organs, frequently accompanied by vascular inflammation. This endothelial syncytia formation has been described as "perhaps the most unique histopathological finding," a feature not described in viral encephalides caused by other virus families. Importantly, in HeV-infected horses, increased numbers of endothelial syncytia have been associated with decreased survival.[34,35] Infection of endothelial cells and subsequent syncytia formation and endothelial damage has been proposed to trigger the thrombosis and necrosis seen in involved vessels. Thus, host tropism and fusogenicity of the viral glycoproteins [8,18,30] are likely to be significant determinants of viral pathogenicity.

Third, in addition to the endothelial and end organ damage, there was extensive lymphoid necrosis and immunologic detection of virus in lymphoid tissues, with lymphocyte apoptosis and detection of multinucleated giant cells in the lymphoid parenchyma and the subcapsular sinus. The authors have proposed that the exposure of the virus to cognate B cells probably occurs first in lymphoid tissues, given the observation that antibodies to the virus appeared in serum before appearing in the CSF. However, the target cells in lymphoid tissues have not been identified; the recent discovery of the NiV receptor(s) may shed light on this critical unanswered question.

Based on their findings, the authors suggested a temporal mode of events. Upon infection, primary replication of the virus occurs in lymphoid tissues (acute lymphoid necrosis), which leads to viremia and secondary replication in endothelial cells. Subsequent syncytia formation is the sentinel pathogenic event that leads to vasculitis, thrombosis, and necrosis and infection of parenchymal cells in various tissues. However, even though NiV antigen can be found in the microvasculature of many tissues, the highest antigen load is found in the brain parenchyma, especially in neurons. Thus, neuronal and endothelial cell tropism is an established hallmark of NiV. The recent discovery of ephrinB2 as the receptor for NiV entry (see below) is remarkably consistent with the known tropism of NiV, as ephrinB2 is critically involved in neurogenesis and angiogenesis, and is highly expressed in neurons and endothelial cells. Ephrin genes are highly conserved, and have been found in all animal species examined.[36] Thus, the conservation of ephrinB2 may also explain the unusually broad tropism of NiV. EphrinBs bind to a large family of ephB receptor tyrosine kinases. It is the cognate bidirectional signaling that occurs during ephrinB–ephB interactions that direct the migration of endothelial cells and neuronal dendrites.[36–40] The interference with these interactions by NiV-G or HeV-G may also have great relevance for the pathogenicity of the virus.

NiV also has many unusual features, some of which undoubtedly contribute to its severe pathogenicity. For example, the NiV P, V, and W proteins, produced from the same P gene by variant RNA editing, have been demonstrated to exhibit anti-interferon functions by a variety of novel mechanisms.[41,42] These anti-interferon functions may dampen the initial innate immune response against this pathogen and contribute to the severe pathogenicity associated with this virus. However, whether the anti-interferon functions of NiV's P, V, and/or W proteins are qualitatively or quantitatively different from other paramyxoviral homologs remains to be seen.

DISCOVERY OF EPHRINB2 AS THE RECEPTOR FOR NIPAH AND HENDRA VIRUS ENTRY

Border and colleagues had previously demonstrated that NiV and HeV likely use the same cellular receptor, as fusion assays showed that NiV and HeV envelope glycoproteins exhibit the same tropism for a variety of cell lines.[8] These initial experiments suggested that the receptor was widely expressed, but not ubiquitous, as many human hematopoietic cell lines were not permissive for NiV and HeV envelope-mediated fusion. However, these initial experiments did identify several cell lines that were nonpermissive for viral fusion. In separate experiments, using a viral overlay technique to identify candidate proteins that may bind to NiV envelope glycoproteins, this group identified a protein candidate around 50 kDa that was bound specifically by viral particles in their viral overlay assay.[43] Then, using microarray analyses to compare a series of permissive and nonpermissive cell lines, Border and colleagues narrowed down a candidate list of putative receptors that fit several criteria (protein size, predicted membrane protein, highly expressed in permissive but not nonpermissive cell lines). Each of these candidate genes was then transfected into a nonpermissive cell line and tested for its ability to confer fusion permissivity by NiV envelope glycoproteins. Only one of the genes, ephrinB2, was able to confer fusion permissivity. EphrinB2 was subsequently confirmed to be a bonafide receptor in their hands as it enabled live NiV entry into otherwise nonpermissive cell lines.[44]

Concurrently, the author's laboratory had also identified ephrinB2 as the bonafide NiV entry receptor by completely independent means.[31] We had first codon-optimized the ectodomain of the NiV attachment protein (G) and fused it to the constant Fc region of human IgG1. This was a necessary technique as paramyxoviral envelope genes are generally not expressed well from standard RNA Pol II-dependent promoters, at least not at levels required for substantive biochemical studies. Indeed, a large number of studies on paramyxoviral envelopes had relied on vaccinia-driven envelope expression. Thus, our studies on the properties of the Nipah envelope proteins can be performed under a less confounding experimental setting, as it has been noted that the use of vaccinia vectors in studies of paramyxovirus fusion necessitates the consideration of vaccinia structural proteins as a confounding factor,[45] although to our knowledge, reports of actual vaccinia-related confounding factors have not been reported. Our codon-optimized immunoadhesin (NiV-G-Fc) had cell –line-binding properties that recapitulated the published cell line tropism of NiV. Preparative immunoprecipitation followed by mass spectrometric analyses (LC-tandem MS/MS) leads to the identification of ephrinB2 as the protein that was specifically bound by the attachment protein of NiV (NiV-G). Several lines of evidence were used to show that ephrinB2 was the receptor for NiV entry: (1) soluble NiV-G binding to a variety of permissive cells was inhibited by soluble ephrinB2 and a variety of soluble ephB receptors, which are cognate endogenous receptors that bind ephrinB2; (2) transfection of ephrinB2 into a variety of nonpermissive cell lines made them permissive for NiV envelope-mediated fusion and entry; and (3) NiV envelope-mediated viral entry into endothelial cells and neurons, two primary cell types known to be infected by NiV *in vivo*, were inhibited by soluble ephrinB2.[31] Significantly, this latter point provided strong evidence that ephrinB2 is likely the *bonafide* cellular receptor used by NiV *in vivo*.

The ability to detect a specific interaction between the recombinant soluble NiV attachment envelope glycoprotein and a soluble recombinant version of its receptor suggests that this interaction can be exploited to screen chemical libraries in an ELISA-based sort of assay to identify leads for small molecule inhibitors of NiV or HeV entry. Since both NiV and HeV seem to interact with the same two conserved surface residues in ephrinB2 (see below), it suggests that lead compounds identified to inhibit NiV, will also likely inhibit HeV.

CONCORDANCE OF EPHRINB2 EXPRESSION PATTERN WITH KNOWN CELLULAR TROPISM OF NIPAH VIRUS

The expression pattern of ephrinB2 is remarkably concordant with the known cellular tropism of NiV, further underscoring that ephrinB2 is a functional receptor for NiV *in vivo*. Much of our knowledge regarding ephrinB2 expression *in situ* comes from LacZ reporter mice where the β-galactosidase gene was knocked into the ephrinB2 locus on one chromosome [46–49] (ephrinB2-null mice are embryonic lethal). The correlations between ephrinB2 expression and NiV tropism *in vivo* are listed; direct quotes are taken from some of these primary references to show the remarkable degree of concordance without paraphrasing on our part:

1. EphrinB2 is expressed in neurons and endothelial cells, and is required for axonal guidance and angiogenesis.[36,37,40]
   - NiV has a tropism for neurons and endothelial cells; NiV antigen is found in neurons and endothelial cells; and endothelial syncytia are a hallmark of NiV infection.[18,30]
2. EphrinB2 also selectively marks smooth muscle surrounding arteries but not veins.[48,49]
   - "Vascular staining for Nipah antigen is mainly found in endothelium, endothelial syncytia and smooth muscle of tunica media of small arteries." (FIG. 6 E, F in ref. 18)
3. EphrinB2 is expressed in placental tissue during normal development in all three trimesters.[50]
   - Evidence of HeV placental transmission to fetus in bats and guinea pigs.[35]
4. EphrinB2 expression in spleen is clearly "limited to the central arteries, arterioles, and capillaries of white pulp, terminating abruptly at the marginal zone between white and red pulp" (where venous sinuses originate).[48]
   - In NiV-infected humans (from autopsy samples): "the spleen showed white pulp depletion and acute necrotizing inflammation in the periarteriolar sheaths; viral antigen staining is also particularly evident in periarteriolar sheaths in the white pulp."[18,30]
5. EphrinB2 expression in the sinusoidal lining of lymph nodes can be found.[46]
   - Syncytia were detected in sinusoidal lining of NiV-infected cats[30] and also in subcapsular sinus of infected humans.[18]

ENVELOPE–RECEPTOR INTERACTIONS AND PATHOGENICITY

Receptor Clustering and Signaling

Both ephrinB2, and its cognate endogenous receptor ephB4, have tyrosine signaling and PDZ-binding motifs in their cytoplasmic domains.[38] "Forward" signaling through ephB4 mediates antiadhesive and repulsive behavior upon contact with ephrinB2-expressing cells, while ephrinB2 "reverse" signaling mediates propulsive adhesion upon contact with ephB4-expressing cells. If NiV-G acts like ephB4 and binds to ephrinB2, but lacks reverse signaling, perhaps only forward propulsion will ensue. We speculate that this might act to recruit more endothelial cells to areas of NiV replication. Indeed, signaling-deficient ephB4 on tumor cells can promote invasion by ephrinB2-expressing endothelial cells.[51] It will be interesting to reexamine pathological specimens for increased angiogenesis in areas of NiV replication. It is also possible that PDZ domain containing proteins and other signaling proteins known to interact with the cytoplasmic domain of ephrinB2 may also play a role in the productive entry of NiV. For example, it has been proposed that PDZ domain containing proteins may bind the cytoplasmic domain of Eph receptor and ephrinB ligands, and stabilize higher-ordered clustering of these receptor ligand pairs into oligomeric arrays.[52] The density of this clustering may modulate the nature and intensity of the forward and reverse signaling.[52] It is known that both HeV-G and NiV-G are tetrameric,[32,53] like the proposed quaternary structures of ephB receptors. Therefore, the clustering of NiV-G with ephrinB2 during the viral entry process may be an essential component of the entry process, and indeed, any putative signals sent through ephrinB2 upon interaction with NiV-G may play a role in viral pathogenesis, as it may allow the cell to become more permissive for viral replication. Since the critical residues in ephrinB2 involved in interactions with NiV-G are the same as those required for interaction with the ephB2 receptor,[33] it is likely that NiV-G will indeed induce reverse signaling via ephrinB2. This is a phenomenon that requires active investigation. Envelope-mediated receptor signaling is an established phenomenon in HIV entry pathogenesis,[54–59] and has even been proposed to play a role in promoting HIV replication in unactivated CD4+ T cells.[60,61]

EphrinB3 as an Alternate Receptor

The virological literature is replete with examples of how minor changes in envelope can lead to attenuation or increased virulence.[62–67] For viruses that can use alternate receptors, which are generally related members of the same family, qualitatively and quantitatively different receptor usage can often account for the different pathogenic profiles. This phenomenon has received much attention in the HIV field where viruses that use CCR5 are generally less cytopathic, and are found earlier on in infection, while viruses that use CXCR4 are generally more cytopathic, and appear with the onset of clinical AIDS.[68–71]

In NiV and HeV infections, despite the similar endothelial cell tropism seen, comparative pathological studies have noted a greater likelihood of meningitis or encephalitic pathology in NiV compared to HeV infections.[30] In pigs, as opposed to most other susceptible animals, the NiV is highly transmissible, with viral antigen readily detectable in respiratory secretions.[29,30] HeV does not appear to be as readily transmissible,[30,35] although admittedly the number of HeV cases is very small.

Can differential receptor usage contribute to the variant pathogenic profiles observed between NiV and HeV disease? In the published literature, ephrinB2 is clearly the primary receptor used by NiV and HeV in a variety of permissive cell lines.[31,44] In the case for NiV, ephrinB2 also appears to be the receptor used in primary neurons and microvascular endothelial cells.[31] There are at least three ephrinB members in the mammalian genome. While ephrinB1 appears not to support NiV-G binding under the *in vitro* conditions examined,[31,33] ephrinB3 appears to be an alternate receptor that can fully support NiV entry, albeit less efficiently than ephrinB2.[33]

Our first hint that an alternate receptor for NiV may exist was our observation that although ephB4, which binds *only* ephrinB2 *in vivo*, was able to block NiV entry into microvascular endothelial cells,[31] it did not appear to block NiV entry into a subset of embryonic rat neurons (unpublished data). Thus, it is likely that while ephrinB2 is the major receptor for NiV entry into microvascular endothelial cells, NiV can also use an alternate receptor for entry into a particular subset of neurons.

We then formally identified ephrinB3 as a *bonafide* receptor for NiV entry using both Nipah envelope pseudotyped VSV reporter viruses, as well as live Nipah viruses.[33] Using biochemical and biophysical measurements, we also show that NiV-G interacts directly with ephrinB3, but with approximately 10-fold lower affinity than ephrinB2. However, due to the extraordinarily high affinity of NiV-G for ephrinB2 (Kd~0.06 nM), the affinity of ephrinB3 was still in the subnanomolar range (Kd~0.6 nM), which may explain why ephrinB3-mediated entry remains readily detectable.[33] In addition, we also pinpointed two key residues common to ephrinB2 and ephrinB3 that mediate their use as receptors for NiV entry. Significantly, these two residues were also critical for their activity as ligands for the endogenous ephB2 receptor (recall that ephrinB ligands bind to ephB receptors). We suggested that ephrinB3 is a relevant receptor for NiV *in vivo*, as MRI lesions in NiV-infected patients can be found in specific regions of the CNS (such as the corpus callosum and the spinal cord)[72,73] that are known only to express ephrinB3, but not ephrinB2.[74,75] Thus, ephrinB3 may serve as a receptor for NiV entry into certain neuronal subsets.

In addition, ephrinB3 expression has been reported in lymphoid cells (T cells),[76] and could also account for acute lymphoid necrosis seen in early NiV infection.[18] While ephrinB2 knockout mice are embryonic lethals with defects in vascular and neuro morphogenesis,[37,40,47–49] ephrinB3 knockout mice are fertile and phenotypically normal, but exhibit peculiar neurological defects arising from the inability of the corticospinal tracts to cross the midline.[77,78] The lack of overt defects in vascular morphogenesis in the ephrinB3 knockout mice suggests that ephrinB3, unlike ephrinB2, is not required for vascular development. Additional support that ephrinB3 is not expressed in endothelium comes from our observation that ephB4, which only binds ephrinB2, can completely inhibit NiV entry into microvascular endothelial cells.[31] Thus, if ephrinB3 is used as an alternate receptor by NiV and/or HeV, the efficiency by which NiV or HeV uses ephrinB2 versus ephrinB3 could contribute to their differential pathologies. Indeed, since NiV appears to have an increased encephalogenic potential compared to HeV, we speculate that NiV may use ephrinB3 better than HeV, as certain neuronal subsets express both ephrinB2 and ephrinB3 while microvascular endothelial cells only express ephrinB2. This hypothesis is readily testable.

N-GLYCANS ON NIPAH VIRUS ENVELOPE CAN BE RECOGNIZED BY IMMUNE SYSTEM LECTINS

Recently, we have also published that galectin-1 can potently inhibit NiV envelope glycoprotein-mediated cell–cell fusion.[32] In exploring the glycan structures present on NiV-F/G, we noticed that both F and G have polylactosamine sequences that can be bound by galectin-1, an immune system lectin that has myriad functions and is known to be a "tuner" of the inflammatory immune response.[79,80] Since gal-1 is produced by activated endothelial cells[81] (a primary target for NiV infection) and dendritic cells (unpublished observations), we asked whether gal-1 could also serve as an innate immune defense mechanism, by recognizing the pathogen glycan structures on NiV-F/G. We found that gal-1 can potently inhibit NiVF/G-mediated cell–cell fusion in a manner that was dependent on its native homodimeric form, and also dependent on its glycan-binding properties.[32] Importantly, we also determined that gal-1 specifically binds NiV-F and -G and indeed, pinpointed an N-linked glycosylation site on NiV-F that is preferentially bound by gal-1: mutation of that N-linked glycosylation site renders the mutant NiV-F significantly more resistant to gal-1-mediated inhibition. This gal-1-mediated inhibition appears to be relatively specific as gal-1 also inhibited HeV-mediated fusion, but not a variety of other viruses tested.[32] Indeed, gal-1 has been recently shown to *enhance* HIV-1 infection.[82] Thus, gal-1 does not appear to have a ubiquitous antiviral effect, unlike cyanovirin, another lectin that has been shown to inhibit a variety of viruses via steric hindrance mechanisms as a result of binding high-mannose glycans present on their envelope glycoproteins.[83,84] The exact mechanism by which gal-1 inhibits NiV fusion remains to be determined, but since gal-1 can aberrantly oligomerize both NiV-F and G,[32] it is not unreasonable to propose that these aberrant oligomers compromise the ability of the F and G complex to undergo the conformational changes required for membrane fusion.

The concentrations of gal-1 used to inhibit NiV fusion can arguably be found in extracellular matrix.[85,86] If this is so, the question remains as to why the natural amount of gal-1 is not more protective against NiV infection. It is difficult to answer this teleological question, but perhaps one could also argue that NiV infection would be even more lethal in the absence of endogenous gal-1. In addition, natural polymorphisms in the gal-1 gene, some of which can be predicted to affect the glycan-binding properties of gal-1, may contribute to the variant mortality rates found between the various NiV outbreaks. However, the frequency of these polymorphisms in the affected populations is not known. In addition, it is also not known what the variance of gal-1 levels is in the general population. Perhaps a threshold level of gal-1 is required for its putative protective effect, and these levels are not present in individuals who succumb to the more severe effects of the Nipah viral infection. Indeed, these questions underscore the need to further understand the potential role that innate immune defense mechanisms may play against emerging pathogens.

CONCLUSION

Nipah virus is an emerging pathogen that is likely a zoonosis that occurred as a result of habitat destruction of its primary host, the fruit bat (*Pteropus hypomelanus* and related species).[87] The rapid population growth in South East Asia and the encroachment of human activity into previously undisturbed areas in the region increases the likelihood that more emerging viruses will be discovered. If we are to better control and prevent any future outbreaks, the biological investigation of any new outbreaks must be coupled with an ecological study of the circumstances and parameters that led to these new zoonoses. Thus, the laudable studies by Daszak, Epstein, and colleagues of the Henipavirus Ecology Research Group deserve much attention and support.[88–90]

ACKNOWLEDGMENTS

This work is supported by NIH grants R21 AI59051, RO1 AI60694, UO1AI70495, and RO1 AI69317. Benhur Lee was a Charles E. Culpepper Medical Scholar supported by the Rockefeller Brothers Fund and was also a recipient of the Burroughs Wellcome Fund Career Development Award. Due to space constraints, the author apologizes for not being able to include all references that might have made this review even more informative.

REFERENCES

1. FIELD, H. *et al.* 2001. The natural history of Hendra and Nipah viruses. Microbes Infect. **3:** 307–314.
2. CHUA, K.B. *et al.* 2000. Nipah virus: a recently emergent deadly paramyxovirus. Science **288:** 1432–1435.

3. CHUA, K.B. *et al.* 1999. Fatal encephalitis due to Nipah virus among pig-farmers in Malaysia. Lancet **354:** 1257–1259.
4. 2004. Nipah virus outbreak(s) in Bangladesh, January-April 2004. Wkly. Epidemiol. Rec. **79:** 168–171.
5. 2005. Emerging infections update: November 2004 to January 2005. CDR Weekly **15:** 6.
6. BUTLER, D. 2004. Fatal fruit bat virus sparks epidemics in southern Asia. Nature **429:** 7.
7. ENSERINK, M. 2004. Emerging infectious diseases. Nipah virus (or a cousin) strikes again. Science **303:** 1121.
8. BOSSART, K.N. *et al.* 2002. Membrane fusion tropism and heterotypic functional activities of the Nipah virus and Hendra virus envelope glycoproteins. J. Virol. **76:** 11186–11198.
9. OLSON, J.G. *et al.* 2002. Antibodies to Nipah-like virus in bats (*Pteropus lylei*), Cambodia. Emerg. Infect. Dis. **8:** 987–988.
10. MACKENZIE, J.S. *et al.* 2001. Emerging viral diseases of Southeast Asia and the Western Pacific. Emerg. Infect. Dis. **7:** 497–504.
11. HSU, V.P. *et al.* 2004. Nipah virus encephalitis reemergence, Bangladesh. Emerg. Infect. Dis. **10:** 2082–2087.
12. TAN, C.T. & K.T. WONG. 2003. Nipah encephalitis outbreak in Malaysia. Ann. Acad. Med. Singapore **32:** 112–117.
13. CHAN, K.P. *et al.* 2002. A survey of Nipah virus infection among various risk groups in Singapore. Epidemiol. Infect. **128:** 93–98.
14. CHUA, K.B. 2003. Nipah virus outbreak in Malaysia. J. Clin. Virol. **26:** 265–275.
15. ICDDRB. 2004. Person-to-person transmission of Nipah virus during outbreak in Faridpur District. Hlth. Sci. Bull. **2:** 5–9.
16. LAM, S.K. 2003. Nipah virus-a potential agent of bioterrorism? Antiviral Res. **57:** 113–119.
17. TAN, C.T. *et al.* 2002. Relapsed and late-onset Nipah encephalitis. Ann. Neurol. **51:** 703–708.
18. WONG, K.T. *et al.* 2002. Nipah virus infection: pathology and pathogenesis of an emerging paramyxoviral zoonosis. Am. J. Pathol. **161:** 2153–2167.
19. USDA. 2002 Summary. Meat animals production, disposition, and income.
20. MOHD NOR, M.N., C.H. GAN & B.L. ONG. 2000. Nipah virus infection of pigs in peninsular Malaysia. Rev. Sci. Tech. **19:** 160–165.
21. BOSSART, K.N. *et al.* 2001. Functional expression and membrane fusion tropism of the envelope glycoproteins of Hendra virus. Virology **290:** 121–135.
22. ALI, R. *et al.* 2001. Nipah virus among military personnel involved in pig culling during an outbreak of encephalitis in Malaysia, 1998–1999. Emerg. Infect. Dis. **7:** 759–761.
23. LAMB, R.A. & D. KOLAKOFSKY. 2001. Paramyxoviridae: the viruses and their replication. *In* Fundamental Virology. D.M. Knipe & P.M. Howley, Eds.: 689–724. Lippincott Williams & Wilkins. Philadelphia.
24. HARCOURT, B.H. *et al.* 2000. Molecular characterization of Nipah virus, a newly emergent paramyxovirus. Virology **271:** 334–349.
25. WANG, L.F. *et al.* 2000. The exceptionally large genome of Hendra virus: support for creation of a new genus within the family Paramyxoviridae. J. Virol. **74:** 9972–9979.
26. SELVEY, L.A. *et al.* 1995. Infection of humans and horses by a newly described morbillivirus. Med. J. Aust. **162:** 642–645.

27. TAMIN, A. *et al.* 2002. Functional properties of the fusion and attachment glycoproteins of Nipah virus. Virology **296:** 190–200.
28. WESTBURY, H. 2000. Hendra virus: a highly lethal zoonotic agent. Vet J. **160:** 165–166.
29. MIDDLETON, D.J. *et al.* 2002. Experimental Nipah virus infection in pigs and cats. J. Comp. Pathol. **126:** 124–136.
30. HOOPER, P. *et al.* 2001. Comparative pathology of the diseases caused by Hendra and Nipah viruses. Microbes Infect. **3:** 315–322.
31. NEGRETE, O.A. *et al.* 2005. EphrinB2 is the entry receptor for Nipah Virus, an emergent deadly paramyxovirus. Nature **436:** 401–405.
32. LEVRONEY, E. *et al.* 2005. Novel innate immune functions for galectin-1: galectin-1 inhibits cell fusion by Nipah virus envelope glycoproteins and augments dendritic cell secretion of proinflammatory cytokines. J. Immunol. **175:** 413–420.
33. NEGRETE, O.A. *et al.* 2006. Two key residues in EphrinB3 are critical for its use as an alternative receptor for Nipah virus. PLoS Pathog. **2:** e7.
34. HOOPER, P.T. *et al.* 1997. Lesions of experimental equine morbillivirus pneumonia in horses. Vet. Pathol. **34:** 312–322.
35. WILLIAMSON, M.M. *et al.* 1998. Transmission studies of Hendra virus (equine morbillivirus) in fruit bats, horses and cats. Aust. Vet. J. **76:** 813–818.
36. POLIAKOV, A., M. COTRINA & D.G. WILKINSON. 2004. Diverse roles of eph receptors and ephrins in the regulation of cell migration and tissue assembly. Dev. Cell. **7:** 465–480.
37. AUGUSTIN, H.G. & Y. REISS. 2003. EphB receptors and ephrinB ligands: regulators of vascular assembly and homeostasis. Cell Tissue Res. **314:** 25–31.
38. KULLANDER, K. & R. KLEIN. 2002. Mechanisms and functions of Eph and ephrin signalling. Nat. Rev. Mol. Cell. Biol. **3:** 475–486.
39. MELLITZER, G., Q. XU & D.G. WILKINSON. 1999. Eph receptors and ephrins restrict cell intermingling and communication. Nature **400:** 77–81.
40. PALMER, A. & R. KLEIN. 2003. Multiple roles of ephrins in morphogenesis, neuronal networking, and brain function. Genes Dev. **17:** 1429–1450.
41. HORVATH, C.M. 2004. Weapons of STAT destruction. Interferon evasion by paramyxovirus V protein. Eur. J. Biochem. **271:** 4621–4628.
42. RODRIGUEZ, J.J. & C.M. HORVATH. 2004. Host evasion by emerging paramyxoviruses: Hendra virus and Nipah virus v proteins inhibit interferon signaling. Viral Immunol. **17:** 210–219.
43. EATON, B.T. *et al.* 2004. Henipaviruses: recent observations on regulation of transcription and the nature of the cell receptor. Arch. Virol. Suppl **18:** 122–131.
44. BONAPARTE, M.I. *et al.* 2005. Ephrin-B2 ligand is a functional receptor for Hendra virus and Nipah virus. Proc. Natl. Acad. Sci. USA **102:** 10652–10657.
45. LAMB, R.A. 1993. Paramyxovirus fusion: a hypothesis for changes. Virology **197:** 1–11.
46. MAKINEN, T. *et al.* 2005. PDZ interaction site in ephrinB2 is required for the remodeling of lymphatic vasculature. Genes Dev. **19:** 397–410.
47. ADAMS, R.H. *et al.* 1999. Roles of ephrinB ligands and EphB receptors in cardiovascular development: demarcation of arterial/venous domains, vascular morphogenesis, and sprouting angiogenesis. Genes Dev. **13:** 295–306.
48. GALE, N.W. *et al.* 2001. Ephrin-B2 selectively marks arterial vessels and neovascularization sites in the adult, with expression in both endothelial and smooth-muscle cells. Dev. Biol. **230:** 151–160.

49. SHIN, D. *et al.* 2001. Expression of ephrinB2 identifies a stable genetic difference between arterial and venous vascular smooth muscle as well as endothelial cells, and marks subsets of microvessels at sites of adult neovascularization. Dev. Biol. **230:** 139–150.
50. GOLDMAN-WOHL, D. *et al.* 2004. Eph and ephrin expression in normal placental development and preeclampsia. Placenta **25:** 623–630.
51. NOREN, N.K. *et al.* 2004. Interplay between EphB4 on tumor cells and vascular ephrin-B2 regulates tumor growth. Proc. Natl. Acad. Sci. USA **101:** 5583–5588.
52. PASQUALE, E.B. 2005. Eph receptor signalling casts a wide net on cell behaviour. Nat. Rev. Mol. Cell. Biol. **6:** 462–475.
53. BOSSART, K.N. *et al.* 2005. Receptor binding, fusion inhibition, and induction of cross-reactive neutralizing antibodies by a soluble G glycoprotein of Hendra virus. J. Virol. **79:** 6690–6702.
54. WEISSMAN, D. *et al.* 1997. Macrophage-tropic HIV and SIV envelope proteins induce a signal through the CCR5 chemokine receptor. Nature **389:** 981–985.
55. DAVIS, C.B. *et al.* 1997. Signal transduction due to HIV-1 envelope interactions with chemokine receptors CXCR4 or CCR5. J. Exp. Med. **186:** 1793–1798.
56. LIU, Q.H. *et al.* 2000. HIV-1 gp120 and chemokines activate ion channels in primary macrophages through CCR5 and CXCR4 stimulation. Proc. Natl. Acad. Sci. USA **97:** 4832–4837.
57. FREEDMAN, B.D. *et al.* 2003. HIV-1 gp120 chemokine receptor-mediated signaling in human macrophages. Immunol. Res. **27:** 261–276.
58. CICALA, C. *et al.* 2006. R5 and X4 HIV envelopes induce distinct gene expression profiles in primary peripheral blood mononuclear cells. Proc. Natl. Acad. Sci. USA **103:** 3746–3751.
59. TOMKOWICZ, B. *et al.* 2006. The Src kinase Lyn is required for CCR5 signaling in response to MIP-1{beta} and R5 HIV-1 gp120 in human macrophages. Blood **108:** 1145–1150.
60. LIN, Y.L. *et al.* 2006. The efficiency of R5 HIV-1 infection is determined by CD4 T-cell surface CCR5 density through Galphai-protein signalling. Aids **20:** 1369–1377.
61. LIN, Y.L. *et al.* 2005. G-protein signaling triggered by R5 human immunodeficiency virus type 1 increases virus replication efficiency in primary T lymphocytes. J. Virol. **79:** 7938–7941.
62. DAL MONTE, P. *et al.* 2004. Genomic variants among human cytomegalovirus (HCMV) clinical isolates: the glycoprotein n (gN) paradigm. Hum. Immunol. **65:** 387–394.
63. WEAVER, S.C. *et al.* 2004. Genetic determinants of Venezuelan equine encephalitis emergence. Arch. Virol. Suppl. **18:** 43–64.
64. CLAPHAM, P.R. & A. MCKNIGHT. 2002. Cell surface receptors, virus entry and tropism of primate lentiviruses. J. Gen. Virol. **83:** 1809–1829.
65. SHUKLA, D. & P.G. SPEAR. 2001. Herpesviruses and heparan sulfate: an intimate relationship in aid of viral entry. J. Clin. Invest. **108:** 503–510.
66. GALLAGHER, T.M. & M.J. BUCHMEIER. 2001. Coronavirus spike proteins in viral entry and pathogenesis. Virology **279:** 371–374.
67. KIDO, H. *et al.* 1999. Cellular proteinases trigger the infectivity of the influenza A and Sendai viruses. Mol. Cells **9:** 235–244.
68. MOORE, J.P. *et al.* 2004. The CCR5 and CXCR4 coreceptors–central to understanding the transmission and pathogenesis of human immunodeficiency virus type 1 infection. AIDS Res. Hum. Retroviruses **20:** 111–126.

69. Ross, T.M., P.D. Bieniasz & B.R. Cullen. 1999. Role of chemokine receptors in HIV-1 infection and pathogenesis. Adv. Virus Res. **52:** 233–267.
70. Berger, E.A., P.M. Murphy & J.M. Farber. 1999. Chemokine receptors as HIV-1 coreceptors: roles in viral entry, tropism, and disease. Annu. Rev. Immunol. **17:** 657–700.
71. Doms, R.W. & S.C. Peiper. 1997. Unwelcomed guests with master keys: how HIV uses chemokine receptors for cellular entry. Virology **235:** 179–190.
72. Lim, C.C. *et al.* 2000. Nipah viral encephalitis or Japanese encephalitis? MR findings in a new zoonotic disease. AJNR Am. J. Neuroradiol. **21:** 455–461.
73. Lim, C.C. *et al.* 2003. Late clinical and magnetic resonance imaging follow up of Nipah virus infection. J. Neurol. Neurosurg. Psychiatry **74:** 131–133.
74. Liebl, D.J. *et al.* 2003. mRNA expression of ephrins and Eph receptor tyrosine kinases in the neonatal and adult mouse central nervous system. J. Neurosci. Res. **71:** 7–22.
75. Benson, M.D. *et al.* 2005. Ephrin-B3 is a myelin-based inhibitor of neurite outgrowth. Proc. Natl. Acad. Sci. USA **102:** 10694–10699.
76. Yu, G. *et al.* 2003. Mouse ephrinB3 augments T-cell signaling and responses to T-cell receptor ligation. J. Biol. Chem. **278:** 47209–47216.
77. Bergemann, A.D. *et al.* 1998. Ephrin-B3, a ligand for the receptor EphB3, expressed at the midline of the developing neural tube. Oncogene **16:** 471–480.
78. Kullander, K. *et al.* 2003. Role of EphA4 and EphrinB3 in local neuronal circuits that control walking. Science **299:** 1889–1892.
79. Rabinovich, G.A. *et al.* 2002. Galectins and their ligands: amplifiers, silencers or tuners of the inflammatory response? Trends Immunol. **23:** 313–320.
80. Almkvist, J. & A. Karlsson. 2004. Galectins as inflammatory mediators. Glycoconj J. **19:** 575–581.
81. Baum, L.G. *et al.* 1995. Synthesis of an endogeneous lectin, galectin-1, by human endothelial cells is up-regulated by endothelial cell activation. Glycoconj J. **12:** 63–68.
82. Ouellet, M. *et al.* 2005. Galectin-1 acts as a soluble host factor that promotes HIV-1 infectivity through stabilization of virus attachment to host cells. J. Immunol. **174:** 4120–4126.
83. Botos, I. & A. Wlodawer. 2003. Cyanovirin-N: a sugar-binding antiviral protein with a new twist. Cell. Mol. Life Sci. **60:** 277–287.
84. Barrientos, L.G. & A.M. Gronenborn. 2005. The highly specific carbohydrate-binding protein cyanovirin-N: structure, anti-HIV/Ebola activity and possibilities for therapy. Mini Rev. Med. Chem. **5:** 21–31.
85. He, J. & L.G. Baum. 2004. Presentation of galectin-1 by extracellular matrix triggers T cell death. J. Biol. Chem. **279:** 4705–4712.
86. Collins, B.E. & J.C. Paulson. 2004. Cell surface biology mediated by low affinity multivalent protein-glycan interactions. Curr. Opin. Chem. Biol. **8:** 617–625.
87. Wong, K.T. *et al.* 2002. Nipah virus infection, an emerging paramyxoviral zoonosis. Springer Semin. Immunopathol. **24:** 215–228.
88. Epstein, J.H. *et al.* 2006. Feral cats and risk for Nipah virus transmission. Emerg. Infect. Dis. **12:** 1178–1179.
89. Epstein, J.H. *et al.* 2006. Nipah virus: impact, origins, and causes of emergence. Curr. Infect. Dis. Rep. **8:** 59–65.
90. Daszak, P. *et al.* 2004. Conservation medicine and a new agenda for emerging diseases. Ann. N. Y. Acad. Sci. **1026:** 1–11.

Avian and Human Metapneumovirus

SHOBHA BROOR AND PREETI BHARAJ

Department of Microbiology, All India Institute of Medical Sciences, New Delhi, India

ABSTRACT: Pneumovirus infection remains a significant problem for both human and veterinary medicine. Both avian pneumovirus (aMPV, Turkey rhinotracheitis virus) and human metapneumovirus (hMPV) are pathogens of birds and humans, which are associated with respiratory tract infections. Based on their different genomic organization and low level of nucleotide (nt) and amino acid (aa) identity with paramyxoviruses in the genus *Pneumovirus*, aMPV and hMPV have been classified into a new genus referred to as *Metapneumovirus*. The advancement of our understanding of pneumovirus biology and pathogenesis of pneumovirus disease in specific natural hosts can provide us with strategies for vaccine formulations and combined antiviral and immunomodulatory therapies.

KEYWORDS: avian pneumovirus; human metapneumovirus; pathogenesis; nucleotide; sequence identity

INTRODUCTION

Avian pneumovirus (aMPV), also called turkey rhinotracheitis virus (TRTV), and a recently discovered human metapneumovirus (hMPV) are classified in the Metapneumovirus genus within the Paramyxoviridae family of viruses. These two viruses are responsible for causing respiratory diseases of turkeys and humans.

The *Paramyxovirinae* and *Pneumovirinae* subfamilies of the *Paramyxoviridae* family include several major pathogens of humans and animals. The subfamily *Pneumovirinae* is taxonomically divided into two genra *Pneumovirus* and *Metapneumovirus*.[1] Human respiratory syncytial virus (hRSV), the type species of the *Pneumovirus* genus, is the single most important cause of acute lower respiratory tract infections (ALRIs) during infancy and early childhood worldwide.[2,3] Other members of the *Pneumovirus* genus include bovine respiratory syncytial virus (bRSV), ovine respiratory syncytial virus (oRSV), and pneumonia virus of mice (PVM). Avian pneumovirus, which causes upper

Address for correspondence: Shobha Broor, Room number 2100, Department of Microbiology, All India Institute of Medical Sciences Hospital, New Delhi, India, 110029. Voice: +919811873410; fax: 01126588663.
shobha.broor@gmail.com

Ann. N.Y. Acad. Sci. 1102: 66–85 (2007). © 2007 New York Academy of Sciences.
doi: 10.1196/annals.1408.005

respiratory tract infection (URI) of turkeys, is included in genus *Metapneumovirus* and is now referred to as avian metapneumovirus (aMPV).[4] In 2001, the isolation of a new paramyxovirus from children with ALRI in the Netherlands was reported, which was tentatively assigned to the *Metapneumovirus* genus of the *Pneumovirinae* subfamily[5] and is now classified as hMPV.

PNEUMOVIRINAE

Electron microscopic analysis of the virions of all the members of the subfamily *Pneumovirinae* including metapneumovirus demonstrates that they are structurally similar to one another and similar to other members of the family *Paramyxoviridae*. The virions are typically spherical (diameters of 150 to 200 nm), with some degree of pleomorphism, and filamentous particles of up to 400 nm in length can be seen.[6,7] The virion is surrounded by a lipid envelope derived from the plasma membrane of the host cell into which the three virus glycoproteins, the attachment (G), fusion (F), and small hydrophobic (SH) proteins, are inserted.[8,9] The F and G glycoproteins, which form the 10–14 nm spikes on the surface of the virus particle, interact with the virus matrix (M) protein layer on the internal surface of the envelope.[10] The details of the molecular interaction(s) between the M protein and the surface F and G glycoprotein(s) are not clear. The helical nucleocapsid is located within the M-protein layer, and includes the 13–15-kb single-stranded, nonsegmented negative sense RNA genome in association with the virus nucleoprotein (N), large (L) polymerase protein, and the phosphoprotein (P). The nucleocapsids of the members of *Pneumovirinae* (diameter 13 to 14 nm) are significantly smaller than *Paramyxovirinae* (18 nm).[11]

Genome Organization of the Pneumovirinae

The genome of the members of subfamily *Pneumovirinae* is a single-stranded negative sense RNA, which varies in size in the two genra. In genus *Pneumovirus,* the size of the genome is approximately 15,000 kb in length compared to genus *Metapneumovirus* wherein genomes are smaller being approximately13,300 kb.[12] The *Pneumovirus* genome encodes 10 transcription units including nonstructural proteins NS1 and NS2 followed by structural proteins, that is, nucleocapsid (N), phosphoprotein (P), matrix (M), small hydrophobic (SH), surface glycoprotein (G), fusion (F), second matrix (M2), and RNA-dependent RNA polymerase (L),[4, 13] whereas the *Metapneumovirus* genome encodes 8 distinct transcription units, barring NS1 and NS2. In contrast to the genomic organization of the *Pneumovirus* being **3′-NS1-NS2-N-P-M-SH-G-F-M2- L-5′-**, *Metapneumovirus* have a different positioning of the genes and the order is **3′–N-P-M-F- M2-SH-G-L-5′-.**[14,15] In pneumovirus F-protein gene follows the G-protein gene, whereas in metapneumovirus the order is reversed, that is, G-protein gene follows fusion, matrix-2, and SH protein genes.[16] Thus, there is absence of ORFs between the M and F genes in hMPV compared to hRSV.[12]

In *Pneumovirinae* the number of nucleotides does not follow the "rule of six"[17–19] compared to *Paramyxovirinae*. The reasons underlying the discrepancy have not been determined, but it has been proposed that length restriction among the *Paramyxovirinae* may be due to specific interactions between N protein and RNA (e.g., binding units of six nucleotides) and/or the need to ensure the integrity of the genome in viruses susceptible to non-template-dependent nucleotide insertion. The genomes of the pneumoviruses and metapneumoviruses encode a greater number of mRNAs (10 and 8, respectively) than the 6 or 7 mRNAs encoded by genomes of the *Paramyxovirinae.*[12] While all the members of the family *Paramyxovirinae* encode a standard set of structural proteins,[16] one feature that differentiates the *Paramyxovirinae* from the *Pneumovirinae* is the different coding potential of P genes and the use of an RNA editing mechanism to express multiple proteins from this gene in *Pneumovirinae.*[20] Another unique feature of the *Pneumovirinae* is the gene encoding the M2 proteins that have not been identified in any *Paramyxovirinae.*[21] The genomes of the *Pneumovirinae* contain regulatory sequences at each terminus. The 5′- end of the genome directs replication and transcription, while the 3′-end contains signals that direct replication of the antigenome for synthesis of progeny virions.[22]

Virus Attachment and Entry

The conventional model of attachment for the members of subfamily *Pneumovirinae* including genus *Pneumovirus* and *Metapneumovirus* involves the interaction of virus G protein with a molecule or molecules on the target cell surface.[23] While the precise nature of the cellular receptor has not been identified, current evidence suggests that one or more cellular glycosaminoglycans or heparin-like molecules are involved.[24–27] Following G-protein-mediated attachment, the F glycoprotein then promotes pH-independent fusion between the cell membrane and the virus envelope. For the F protein to become functional, it has to be cleaved by cellular proteases, and the hydrophobic amino-terminal region of the F_1 component then promotes the fusion process, which introduces the internal components of the virion into the cytoplasm of the host cell, where the remainder of the infectious cycle takes place. While *Pneumovirus* infection results in the alteration of host cell gene expression, there is no absolute requirement for a functional nucleus, since virus development can proceed in enucleated cells.[28]

PNEUMOVIRUS GENE PRODUCTS

Nonstructural (NS1 and NS2) Proteins

The NS1 and NS2 proteins are unique to the *Pneumovirus*. The genes are located at the 3′-end of the coding region of the genome and, as a consequence, are the most abundantly transcribed genes. These proteins are small in size and probably have a regulatory role for RNA synthesis[29] as well as inhibit responses to interferon.[30–32] Metapneumoviruses on the other hand lack NS1 and NS2 genes.[33]

Nucleocapsid (N) Protein

The N protein is an integral part of the nucleocapsid complex of the virion and is an essential component of the polymerase complex. The N protein gives the RNA genome its helical structure through a tight association with the virus genome RNA. The hRSV N protein is 391 amino acids (aa) in length and is smaller in size compared to N protein of the members of *Paramyxovirinae* (489–553 aa).[34] It has a relatively high level of aa identity (60%) to the N protein of PVM[34] and has regions of specific sequence homology to the N proteins of aMPV and hMPV.[35] The hMPV N gene is about 1185 bases in length and the deduced N protein is 394 aa in length with an estimated molecular mass of 43.5 kDa, which is similar in size to the members of genus *Pneumovirus*.[36] The hMPV N protein shows 69–89% aa identity with aMPV and 42–44% with hRSV.[34,35,37,38]

Phosphoprotein (P)

The P protein is an essential component of the replication and transcription complexes of the Pneumoviruses.[39–41] Carboxyl terminus of the P protein contains the elements necessary to bind to the N protein.[42–44] The P-protein genes of the *Pneumovirus* encode several different proteins using alternative start codons. In addition to the full-length sequence, the P gene of hRSV encodes an mRNA for an additional, short polypeptide of unknown function.[45] No additional proteins have been identified from the P genes of the *Metapneumovirus*. The coding region of the P gene in hMPV is 885 bases in length, encoding a predicted protein of 294 aa residues with a molecular mass of 32.5 kDa.[36,46] The size of the hMPV P protein is comparable to those of aMPV and PVM (279–295 aa).[37–39] The P protein of hMPV has a significant level of divergence with other Pneumoviruses.[38–40] The highest levels of aa identity were observed with aMPV (50–68%), with the greatest identity shared with aMPV C (68%).[37] The aa identity level is only 23–28% with mammalian pneumoviruses.[38,40]

Matrix (M) Protein

The M protein is associated with the inner surface of the membrane of infected cells[47,48] and has a carboxyl-terminal hydrophobic domain. The M proteins of Pneumoviruses render the nucleocapsid transcriptionally inactive before packaging and also promote the association of the nucleocapsid complex with the nascent envelope.[49] M gene of hMPV is 765 nucleotides in length, encoding a predicted protein of 254 aa residues with a predicted molecular mass of 27.6 kDa.[36] The levels of aa identity of hMPV are highest with aMPV C (88%) whereas with other pneumoviruses the identity levels are as low as 37–40%.[33,38,40,50]

SH Protein

The SH proteins of Pneumoviruses are small integral membrane proteins. The sizes of the proteins vary considerably, with hRSV having the smallest SH protein (64 aa) and hMPV having the largest (183 aa).[50,51] The precise function of the SH protein is unknown, and analysis of hRSV mutants lacking the SH and/or G genes suggests that the SH protein is not absolutely necessary for attachment, infectivity, or virion assembly. However, the hRSV G and SH proteins have been implicated in impairing the Th1-mediated host antiviral responses.[52]

Attachment (G) Protein

The *Pneumovirinae* G protein genes do not show any sequence or structural homology to HN or H attachment protein genes of the members of subfamily *Paramyxovirinae*. This gene is the most variable of all the *Pneumovirinae* proteins. The variability in G gene is used for phylogenetic classification of viruses within a species. The G proteins of *Pneumovirus* are heavily glycosylated with both N- and O-glycosylated sites, adding approximately 55 kDa to the mass of the hRSV polypeptide.[53] While the roles of these glycosyl groups have not been completely clarified, extensive glycosylation of the extracellular domain of the G protein could reduce its antigenicity by shielding the virus protein with host-specified sugars. Alternatively, partial utilization of glycosylation sites can result in antigenic heterogeneity, representing a source of diversity within a genetically homogenous virus population.[53,54]

The G gene of aMPV C is 1321 bases in length with one ORF coding for a protein of 435 aa. The structural characteristics of G protein of aMPV C is similar to hRSV G protein including two mucin-like regions The aMPV C, G gene is closer to hMPV compared to aMPV A, B, or D.[55]

Fusion (F) Protein

The F protein of pneumoviruses shows only limited aa identity to F proteins of viruses of family *Paramyxovirinae* but structurally they are similar. The F protein of hRSV is synthesized as an inactive precursor (F_0) that assembles into a homo-oligomer in the rough endoplasmic reticulum (ER), forming a structure that is presumed to represent a single virion spike.[56] When the F_0 protein in trans Golgi network is activated by cleavage into two disulfide-linked subunits, F_1 and F_2, at two furin consensus sites,[57–59] the short (27-aa) peptide between the two furin cleavage sites does not remain associated with the F-protein subunits.[60] The F gene of hMPV is 1620 bases in length, and encodes for a protein of 539 aa in length with a predicted molecular mass of 58.4 kDa. The levels of aa identity with aMPV are 67–82% and with mammalian pneumovirus 32–43%.[36]

M2 Proteins

The genes encoding the M2 proteins of genus *Pneumovirus* contain two overlapping open reading frames.[15,21,61] The first open reading frame, initiated by the 5′–proximal AUG initiation codon, encodes the M2-1 protein, which is involved in virus RNA synthesis.[56] The data suggest that the M2-1 protein functions as a transcription elongation factor.[62] The product of the second open reading frame, the M2-2 protein, has been detected in cells infected with hRSV and PVM.[21] No M2-2 protein was detected in aMPV-infected cells, and to date there is no information about the putative M2-2 protein of hMPV. Expression of the M2-2 open reading frame results in the inhibition of virus gene expression from synthetic genomes. The hRSV M2-2 protein accumulates during infection, and it has been proposed that the role of this protein is to switch the virus from replicative-to-assembly mode prior to virion release.[11,21]

Large (L) Polymerase Protein

The pneumovirus L protein is thought to be the major component of the viral RNA-dependent RNA polymerase complex, which is responsible for the synthesis of all viral RNA, including mRNA, replicative intermediates, and the progeny RNA genomes. The L protein is also thought to be responsible for mRNA methylation and capping. The L protein is very large—the hRSV L protein contains 2,165 aa, and the hMPV L protein contains 2,000 aa.[33,41,63] Six conserved domains for polymerase proteins of minus-strand RNA viruses have been described,[64,65] including three functional domains and distinct interdomain regions acting as hinges to assemble the functional domains into an appropriate conformation.

The intergenic regions play an important role in the regulation of transcription in pneumoviruses. These regions vary 2–56 nucleotides in length and show extensive genomic variability.

AVIAN METAPNEUMOVIRUS

aMPV, like other members of *Pneumovirinae*, is an enveloped nonsegmented negative-stranded RNA virus. The avian metapneumovirus has been classified into four subtypes A, B, C, and D based on nucleotide (nt) and predicted aa sequence analyses.[55] Nucleotide and predicted aa homology within subtypes is 97–99%, whereas strains across subtypes have 56–61% nt and 33.2–38% predicted aa sequence identities within the highly variable attachment G gene.[66] European viruses are mainly subtype A and B whereas in U.S. subtype C viruses have been identified. Subtype D was reported from France.[67] Phylogenetic comparison of subgroups A, B, and C viruses indicated that A and B viruses are more closely related to each other than either A or B viruses were to C viruses. The F and M genes of Colorado aMPV isolate showed 60% nt and 78% aa sequence identity with subgroup A and B viruses whereas between subgroup A and B viruses in M and F genes there is 83–89% predicted aa sequence identity.[4] The analysis of entire gene sequence of aMPVs has shown that aMPV C has a closer relationship with hMPV compared to aMPV subtypes A, B, and D.[4,5,55]

Epidemiology

aMPV was first isolated from turkeys in South Africa during the late 1970s.[68–70] Thereafter, the virus was reported in France,[71] UK,[72] Germany,[73] the Netherlands,[74] Japan,[75] Mexico,[76] Israel,[77] Morocco,[78] Brazil,[79] South America,[80] and then in the United States.[81,82] The disease appeared in UK during the mid-1980s and spread rapidly, affecting turkey flocks of all ages throughout England and Wales.[73,83] The United States was free from aMPV until the first outbreak occurred among commercial turkeys in the state of Colorado in 1996.[82,84] In the subsequent years (1997–2002), aMPV disease emerged as a major economic problem for turkey farmers in Minnesota, which is the largest turkey producing state in the United States. The U.S. turkey outbreaks of aMPV display a seasonal pattern with approximately 80% of the outbreaks occurring in spring (April to May) and autumn (October to December), whereas in UK seasonal pattern of infection is not seen.[80,85,86] aMPV has also been isolated from wild Canada geese and from sentinel ducks placed in a pond in close proximity with an aMPV-infected turkey farm. Sequencing data from the N, P, M, F, and M2 genes from sentinel duck aMPV isolate shared 95–99% nucleotide identity and 97–99% aa identity with four aMPV isolates from a neighboring turkey farm.[86] Whereas these findings clearly demonstrate that aMPV can replicate in wild birds, the epidemiologic picture of the U.S. outbreaks has made it difficult to ascribe a major role to migratory birds in disseminating the virus. The reason for this is because Canada, which neighbors Minnesota to the north, and U.S. states neighboring Minnesota (Iowa, Illinois, South Dakota, North Dakota, and Wisconsin) have not reported serious aMPV outbreaks,[85] in spite of the high incidence of the disease in Minnesota.

Clinical Features in Birds

The *Turkey rhinotracheitis virus*, commonly referred to as aMPV, causes acute rhinotracheitis, which is characterized by coughing, nasal and ocular discharge, swelling of infraorbital sinuses, tracheal rales, and foamy conjunctivitis, in turkeys of all ages. In laying birds, there is a transient drop in egg production, along with mild respiratory tract illness. In infected flocks, there is a high morbidity varying from 50–100% and mortality of 0–30%, which can rise to 90% in conjunction with secondary bacterial infection.[87]

Avian Metapneumovirus in Other Birds

aMPV subtypes A and B can also infect chickens resulting in a condition referred to as swollen head syndrome.[70,72] In addition, aMPV can infect ducks,[88] pheasants,[89,90] guinea fowl,[91] and ostriches.[92]

Diagnosis

Virus isolation can be done in tissue culture. Vero cells are generally used to isolate this virus. Cytopathic effect appears after 2 to 3 days. Indirect immunofluorescence staining (IFA) confirms the presence of aMPV indicated by light green fluorescence that imparts a granular appearance to the cytoplasm of cells containing aMPV. Enzyme-linked immunosorbent assay (ELISA) is done in serum samples from turkeys to test for the presence of antibody to aMPV using anti-turkey IgG conjugate as the secondary antibody. In clinical samples, viral RNA can be detected by conserved F gene reverse transcription polymerase chain reaction (RT-PCR).[93]

Conclusion

aMPV is thus an important pathogen of turkeys and other birds in Europe and the United States. Reports from Southeast Asia are rather few. Further studies in Southeast Asia are needed to study the epidemiology of this virus in birds. As these countries are a major mixing vessel for many respiratory viruses like influenza, it is important to establish the link between aMPV C and hMPV though cross-infection between the two species has not been demonstrated.

HMPV

Recently discovered human metapneumovirus (hMPV) was first identified in 2001 by van den Hoogen *et al.* in the Netherlands.[5] The virus was isolated from nasopharyngeal aspirates taken during a 10-year period from 28 hospitalized children and infants with respiratory tract infections who had signs and symptoms similar to those of hRSV infection. The virus produced cytopathic effect in tertiary monkey kidney (tMK) cells similar to that of hRSV, and electron microscopy of supernatant showed the presence of paramyxovirus-like particles. By sequence analysis and genomic organization, the virus was characterized as a member of the genus *Metapneumovirus* of the family *Paramyxoviridae*, of which the only other known member is the avian metapneumovirus (aMPV), the causative agent of an upper respiratory tract infection in turkeys. In the 4 years since its discovery, the virus has been characterized further and its epidemiology and clinical features have been described in many studies world over.

Virology

On the basis of gene arrangement and sequence homology with aMPV, hMPV has been classified as the first human virus in the genus *Metapneumovirus*, subfamily *Pneumovirinae.*[5] hMPV is frequently compared with the only other human pneumovirus hRSV with which it shares many features including electron microscopic appearance, similar cytopathic effect, and fragility. However, hMPV grows slowly in cell culture and requires trypsin for efficient replication.

Heterogeneity of hMPV

Early sequencing data from the first hMPV isolates from the Netherlands indicated high levels of nucleotide identity and a genomic organization similar to aMPV, particularly the absence of the NS genes at the 3′-end of the genome.[5] Recent complete genome sequencing of the hMPV Canadian strain enabled detailed comparisons between the four aMPV subgroups A, B, C, and D and hMPV. There is 56% and 88% aa sequence identity between aMPV and hMPV in nine open reading frames and some sequence identity within the noncoding regions as well.[41] Comparison of N, P, M, F, and M2 proteins of hMPV has shown the highest (overall 80%) aa identity with aMPV C.[3] The close relationship between hMPV and aMPV C has led to a hypothesis that either of the two viruses can cross-infect, that is, that hMPV might infect turkeys and/or aMPV C may infect humans. The restricted host range of hMPV might be because the attachment G gene has low aa sequence identity with aMPV C. aMPV has been detected in the Netherlands in 1992 and unpublished reports indicate that several turkey flocks in Canada are positive for aMPV.[94] However, aMPV has not been reported in Australia whereas hMPV has been isolated there.[95] Thus, as the geographical distribution of aMPV and hMPV is not the same, these two viruses may have originated and spread independent of each other or in countries like Southeast Asia they may have coexisted together and then evolved independent of each other.

Within hMPV, initially two genetic lineages were reported by van den Hoogen and his colleagues.[4] Later several groups around the world, by phylogenetic analysis of the M, N, P, F, and L genes, showed that hMPV isolates were divided into two genetic clusters tentatively named A and B.[36,96,97] Each cluster then has sublineages 1 and 2 as defined by phylogenetic analysis of sequences of F protein and G protein. The fusion protein has approximately 95% aa sequence identity between viruses from the two lineages, whereas the G protein shows only 30% identity.[97] This classification has implications for serologic diagnosis and development of primers for RT-PCR. Recently, the complete consensus nucleotide sequences of isolates representative of both genotypes revealed an overall aa sequence identity of 90% and the greatest diversity was for G gene (37% identity).[98] Furthermore, analysis of several hMPV G ORFs revealed aa sequence identities ranging from 31% to 35% between the two clusters.[99] The nucleotide and aa identity of N, F, P, and M protein between the two clusters and sublineages is shown in TABLE 1.[36]

More recently, a study involving phylogenetic analysis of both the F and the N genes from Germany over two seasons showed a further bipartition of subgroup A2 to tentatively assigned A2a and A2b. There was co-circulation of all described genotypes with subtype A predominating over subtype B, thus confirming previous reports. The idea is that novel sublineages are not locally or temporarily restricted but might be prevalent worldwide.[100]

Epidemiology

Since hMPV has been recently discovered, the complete epidemiology and clinical features of this infection have not been fully described. Infection with hMPV has been described from many countries, including Canada, Brazil, Japan, South Africa, Italy, Australia, United States, Hong Kong, and UK, thus suggesting its worldwide distribution.[96,101–107] The frequency of hMPV detection in patients with respiratory tract infections ranges from 2% to 25%.[105,107]

TABLE 1. Nucleotide and predicted amino acid sequence identity levels within and between the 2 hMPV genetic clusters (Bastien *et al.*[36])

Clusters	N	P	M	F
A/A(nt)	93.8–99.8%	91.0–99.9%	94.1–99.9%	94.3–99.9%
(aa)	(99.3–100%)	(95.0–100%)	(98.8–100%)	(98.3–99.8%)
B/B(nt)	94.5–100%	93.3–100%	94.1–100%	94.4–100%
(aa)	(98.3–100%)	(94.6–100%)	(99.6–100%)	(98.9–100%)
A/B(nt)	85.3–86.1%	78.0–79.4%	83.5–85.6%	83.0–83.6%
(aa)	(94.2–95.9%)	(81.6–85.7%)	(96.9–97.6%)	(94.1–95.4%)

Recently, in a study from India, hMPV has been detected in 12% to 19% of hospital children with mild and severe respiratory tract infection.[108,109] Most hMPV infections occur in < 5 years of age with peak at 4–6 months of age. However, hMPV infections tend to occur in slightly older children compared to hRSV.[97,110] Like hRSV and influenza viruses hMPV has also been described as a significant cause of ARI in the elderly and patients with underlying diseases.[110–112] One study among the elderly indicated that hMPV caused more severe disease in frail elderly patients, such as nursing home residents, than in elderly or younger adults who were fit.[112] Several community studies in adults have helped to define the role of hMPV as the cause of ambulatory ARIs. Two longitudinal studies, one conducted in British general practices over a single winter and the other in adults in Rochester, New York, over two winter seasons, reported hMPV infection rates of 1.3% and 1.5–7.0%, respectively.[107,112] Like hRSV and influenza hMPV also shows seasonal peaks. In year-round surveillance studies, hMPV was found to peak from December to February.[105,111] In Hong Kong, hMPV was found to peak in spring and summer, however, in the temperate regions, hMPV is detected mainly from October to May.[110,111,113]

Clinical Features in Children

Wide spectrums of clinical manifestations ranging from mild URI to severe ALRI have been reported in all age groups with hMPV infection. In a study conducted on more than 2,000 children at Vanderbilt University Medical Center in Nashville, Tennessee, it was observed that hMPV was associated with common cold complicated with otitis media in about one-third of children. In addition, it caused bronchiolitis (59%), pneumonia (8%), croup (18%), and exacerbation of reactive airways disease (14%).[114] The clinical profile of hMPV infection is similar to that of hRSV with bronchiolitis being the most frequent manifestation, followed by bronchopneumonia, pneumonitis, wheezing, and otitis media. Most lower respiratory tract infections are seen in children younger than 1 year of age who are otherwise healthy. In hospitalized children, patients with underlying disease, immunocompromised individuals, and fragile elderly, hMPV disease tends to be more severe. Compared with hRSV infections, the affected children are somewhat older, and the severity of disease is usually somewhat less.[96,101,113,115]

Clinical Features in Adults

In the majority of healthy young and elderly adults, hMPV was predominately a mild disease characterized by cough, hoarseness, congestion, and rhinorrhea.[112] Sore throat, hoarseness, sputum production, and constitutional symptoms were present in more than half of patients.

Human hMPV in Immunocompromised Persons

Although the fatality rate of hMPV appears to be very low, hMPV can be responsible for fatal respiratory insufficiency in severely immunocompromised persons. One young child with acute leukemia was described who suffered from hMPV infection in two subsequent winters. The two strains that were recovered during these two episodes were genetically distinct, and the last episode was fatal.[106] Similarly, a fatal case of hMPV infection was described in an adult hematopoietic stem cell transplant recipient.[116] The role of virus infections in exacerbation of asthma is an issue that needs further studies. Several studies have shown a possible association between hMPV infection and asthma.[111,113,117,118] There is no question that some patients with asthmatic exacerbations have hMPV infection, but whether or not the virus is associated with asthmatic exacerbation more frequently than other viruses (e.g., rhinoviruses) is not yet clear. Larger studies are needed to address this question.

Further co-infection with hMPV and hRSV can occur frequently. In a study by Greensill *et al.*,[119] 70% of children having severe hRSV-related bronchiolitis had hMPV co-infection. Confirmation of a role for co-infection will require longitudinal studies over several bronchiolitis seasons, comparing disease severity with frequency of co-infection across the full spectrum of hRSV disease. This high hRSV–hMPV co-infection rate in severe disease means that previous studies into hRSV subtype/genotype, disease severity, and cytokine and chemokine expression in hRSV disease may need to be reevaluated.

Diagnosis

The virus can be isolated by cell culture. Originally, the virus was isolated in cultures of tMK cells, displaying cytopathological effects within 10 to 14 days postinoculation similar to those seen with hRSV.[5] Boivin *et al.*[101] showed CPE in LLC-MK2 cells after a mean incubation time of 17.3 days, without large syncytia formation. Four principal methods are used for the diagnosis of respiratory virus infections: virus isolation by culture, antigen detection, RNA or DNA detection, and serological study. In clinical samples, viral RNA can be detected by RT-PCR. Several targets for amplification have been chosen in the design of the RT-PCR, and it has been suggested that amplifying the N and/or the L gene is particularly suitable for hMPV diagnosis.[111,120] Since several laboratories have started to implement PCR as routine diagnostic assay for respiratory virus infections, it can be expected that detection of hMPV RNA will be more widely used as part of a respiratory virus diagnosis package. Antibodies against hMPV can be measured and serology studies have been performed, but since everyone over the age of 5 years has anti-hMPV antibodies, antibody detection is not currently implemented as a standard assay in most routine laboratories.[121]

Conclusion

In the 4 years since its discovery, the clinical characteristics of this new virus have been clarified. In children, especially those younger than 1 year of age, hMPV is responsible for a substantial number of respiratory tract infections requiring hospitalization; its clinical course is somewhat milder, but otherwise indistinguishable from hRSV infection. hMPV can also be found in adults, in influenza-like illnesses, but also as a cause of pneumonia. Especially in the latter cases immunosuppressive conditions may be present, and, like hRSV, hMPV can be responsible for respiratory insufficiency under these conditions. The incidence of hMPV as a cause of respiratory failure in these patients needs further investigation.

REFERENCES

1. Virus Taxonomy 7th Report of the International Committee on Taxonomy of Viruses. 2000. M.H. van Regenmortel, C.M. Fauquet & D.H. Bishop, Eds.: 551, 557, 559–560. Academic. San Diego.
2. Domachowske, J.B. & H.F. Rosenberg. 1999. Respiratory syncytial virus infection: immune response, immunopathogenesis, and treatment. Clin. Microbiol. Rev. **12:** 298–309.
3. Selwyn, B.J. 1990. The epidemiology of acute respiratory tract infection in young children: comparison of findings from several developing countries. Rev. Infect. Dis. **12:** S870–S888.
4. Njenga, M.K., H.M. Lwamba & B.S. Seal. 2003. Metapneumoviruses in birds and humans. Virus Res. **91:** 163–169.
5. van den Hoogen, B.G., J.C. de Jong, J. Groen, *et al.* 2001. A newly discovered human pneumovirus isolated from young children with respiratory tract disease. Nat. Med. **7:** 719–724.

6. BACHI, T. & C. HOWE. 1973. Morphogenesis and ultra structure of respiratory syncytial virus. J. Virol. **12:** 1173–1180.
7. JONCAS, J., L. BERTHIAUME & V. PAVILANIS. 1969. The structure of the respiratory syncytial virus. Virology **38:** 493–496.
8. COLLINS, P. L. & G. MOTTET. 1993. Membrane orientation and oligomerization of the small hydrophobic protein of human respiratory Syncytial virus. J. Gen. Virol. **74:** 1445–1450.
9. HUANG, Y.T., P.L. COLLINS & G.W. WERTZ. 1985. Characterization of the ten proteins of human respiratory syncytial virus; identification of a fourth envelope associated protein. Virus Res. **2:** 157–173.
10. COMPANS, R.W., D.H. HARTER & P.W. CHOPPIN. 1967. Studies of pneumonia virus of mice (PVM) in cell culture. II. Structure and morphogenesis of the virus particle. J. Exp. Med. **126:** 267–277.
11. BERTHIAUME, L., J. JONCAS & V. PAVILANIS. 1974. Comparative structure, morphogenesis and biological characteristics of the respiratory syncytial (RS) virus and the pneumonia virus of mice (PVM). Arch. Gesamte Virusforsch. **45:** 39–51.
12. EASTON, A.J., J.B. DOMACHOWSKE & H.F. ROSENBERG. 2004. Animal pneumovirus: molecular genetics and pathogenesis. Clin. Microbiol. Rev. **17:** 390–412.
13. EVANS, J.E., P.A. CANE & C.R. PRINGLE. 1996. Expression and characterization of the NS1 and NS2 proteins of respiratory syncytial virus. Virus Res. **43:** 155–161.
14. LING, R., A.J. EASTON & C.R. PRINGLE. 1992. Sequence analysis of the 22K, SH and G genes of turkey rhinotracheitis virus and their intergenic regions reveal a gene order difference from that of other pneumoviruses. J. Gen. Virol. **73:** 1709–1715.
15. YU, Q., P.J. DAVIS & T.D. BROWN. 1992. Sequence and *in vitro* expression of the M2 gene of turkey rhinotracheitis pneumovirus. J. Gen. Virol. **73:** 1355–1363.
16. PRINGLE, C.R. & A.J. EASTON. 1997. Monopartite negative strand RNA genomes. Semin. Virol. **8:** 49–57
17. BUKREYEV, A., B.R. MURPHY & P.L. COLLINS. 2000. Respiratory syncytial virus can tolerate an intergenic sequence of at least 160 nucleotides with little effect on transcription or replication *in vitro* and *in vivo*. J. Virol. **74:** 11017–11026.
18. CALAIN, P. & L. ROUX. 1993. The rule of six, a basic feature or efficient replication of Sendai virus defective interfering RNA. J. Virol. **67:** 4822–4830.
19. SAMAL, S.K. & P.L. COLLINS. 1996. RNA replication by a respiratory syncytial virus RNA analog does not obey the rule of six and retains a nonviral trinucleotide extension at the leader end. J. Virol. **70:** 5075–5082.
20. CURRAN, J. & D. KOLAKOFSKY. 1999. Replication of paramyxoviruses. Adv. Virus Res. **54:** 403–422.
21. AHMADIAN, G., P. CHAMBERS & A.J. EASTON. 1999. Detection and characterization of proteins encoded by the second ORF of the M2 gene of pneumoviruses. J. Gen. Virol. **80:** 2011–2016.
22. MINK, M.A., D.S. STEC & P.L. COLLINS. 1991. Nucleotide sequences of the 3′ leader and 5′ trailer regions of human respiratory syncytial virus genomic RNA. Virology **185:** 615–624.
23. LEVINE, S., R. KLAIBER-FRANCO & P.R. PARADISO. 1987. Demonstration that glycoprotein G is the attachment protein of respiratory syncytial virus. J. Gen. Virol. **68:** 2521–2524.

24. FELDMAN, S.A., A. AUDET & J.A. BEELER. 2000. The fusion glycoprotein of human respiratory syncytial virus facilitates virus attachment and infectivity via an interaction with cellular heparan sulfate. J. Virol. **74:** 6442–6447.
25. HALLAK, L.K., P.L. COLLINS, W. KNUDSON, *et al.* 2000a. Iduronic acid containing glycosaminoglycans on target cells are required for efficient respiratory syncytial virus infection. Virology **271:** 264–275.
26. HALLAK, L.K., D. SPILLMANN, P.L. COLLINS, *et al.* 2000b. Glycosaminoglycan sulfation requirements for respiratory syncytial virus infection. J. Virol. **74:** 10508–10513.
27. KRUSAT, T. & H.J. STRECKERT. 1997. Heparin-dependent attachment of respiratory syncytial virus (RSV) to host cells. Arch. Virol. **142:** 1247–1254.
28. FOLLETT, E.A., C.R. PRINGLE & T.H. PENNINGTON. 1975. Virus development in enucleate cells: echovirus, poliovirus, pseudorabies virus, reovirus, respiratory syncytial virus and Semliki Forest virus. J. Gen. Virol. **26:** 183–196.
29. ATREYA, P.L., M.E. PEEPLES & P.L. COLLINS. 1998. The NS1 protein of human respiratory syncytial virus is a potent inhibitor of minigenome transcription and RNA replication. J. Virol. **72:** 1452–1461.
30. BOSSERT, B. & K.K. CONZELMANN. 2002. Respiratory syncytial virus (RSV) nonstructural (NS) proteins as host range determinants: a chimeric bovine RSV with NS genes from human RSV is attenuated in interferon-competent bovine cells. J. Virol. **76:** 4287–4293.
31. BOSSERT, B., S. MAROZIN & K.K. CONZELMANN. 2003. Nonstructural proteins NS1 and NS2 of bovine respiratory syncytial virus block activation of interferon regulatory factor-3. J. Virol. **77:** 8661–8668.
32. SCHLENDER, J., B. BOSSERT, U. BUCHHOLZ, *et al.* 2000. Bovine respiratory syncytial virus nonstructural proteins NS1 and NS2 cooperatively antagonize alpha/beta interferon-induced antiviral response. J. Virol. **74:** 8234–8242.
33. RANDHAWA, J.S., A.C. MARRIOTT, C.R. PRINGLE, *et al.* 1997. Rescue of synthetic minireplicons establishes the absence of the NS1 and NS2 genes from avian pneumovirus. J. Virol. **71:** 9849–9854.
34. BARR, J., P. CHAMBERS, C.R. PRINGLE, *et al.* 1991. Sequence of the major nucleocapsid protein gene of pneumonia virus of mice: sequence comparisons suggest structural homology between nucleocapsid proteins of pneumoviruses, paramyxoviruses, rhabdoviruses and filoviruses. J. Gen. Virol. **72:** 677–685.
35. LI, J., R. LING, J.S. RANDHAWA, *et al.* 1996. Sequence of the nucleocapsid protein gene of subgroup A and B avian pneumoviruses. Virus Res. **41:** 185–191.
36. BASTIEN, N., S.W. NORMAND, T.T. TAYLOR, *et al.* 2003. Sequence analysis of the N, P, M and F genes of Canadian human metapneumovirus strains. Virus Res. **93:** 51–62.
37. DAR, A.M., S. MUNIR, S.M. GOYAL, *et al.* 2001. Sequence analysis of the nucleocapsid and phosphoprotein genes of avian pneumoviruses circulating in the US. Virus Res. **79:** 15–25.
38. KARRON, R.A., D.A. BUONAGURIO, A.F. GEORGIU, *et al.* 1997. Respiratory syncytial virus (RSV) SH and G proteins are not essential for viral replication *in vitro*: clinical evaluation and molecular characterization of a cold-passaged, attenuated RSV subgroup B mutant. Proc. Natl. Acad. Sci. USA **94:** 13961–13966.
39. BARR, J., P. CHAMBERS, P. HARRIOTT, *et al.* 1994. Sequence of the phosphoprotein gene of pneumonia virus of mice: expression of multiple proteins from overlapping reading frames. J. Virol. **68:** 5330–5334.

40. SATAKE, M., N. ELANGO & S. VENKATESAN. 1984. Sequence analysis of the respiratory syncytial virus phosphoprotein gene. J. Virol. **52:** 991–994.
41. VAN DEN HOOGEN, B.G., T.M. BESTEBROER, A.D. OSTERHAUS, *et al.* 2002. Analysis of the genomic sequence of a human metapneumovirus. Virology **295:** 119–132.
42. GARCIA-BARRENO, B., T. DELGADO & J.A. MELERO. 1996. Identification of protein regions involved in the interaction of human respiratory syncytial virus phosphoprotein and nucleoprotein: significance for nucleocapsid assembly and formation of cytoplasmic inclusions. J. Virol. **70:** 801–808.
43. HENGST, U. & P. KIEFER. 2000. Domains of human respiratory syncytial virus P protein essential for homodimerisation and for binding to N and NS1 protein. Virus Genes **20:** 221–225.
44. SLACK, M.S. & A.J. EASTON. 1998. Characterisation of the interaction of the human respiratory syncytial virus phosphoprotein and nucleocapsid protein using the two-hybrid system. Virus Res. **55:** 167–176.
45. CARAVOKYRI, C., A.J. ZAJAC & C.R. PRINGLE. 1992. Assignment of mutant tsN19 (complementation group E) of respiratory syncytial virus to the P protein gene. J. Gen. Virol. **73:** 865–873.
46. LAMBDEN, P.R. 1985. Nucleotide sequence of the respiratory syncytial virus phosphoprotein gene. J. Gen. Virol. **66:** 1607–1612.
47. PEEPLES, M.E. & S. LEVINE. 1997. Respiratory syncytial virus polypeptides: their location in the virion. Virology **95:** 137–145.
48. WUNNER, W.H. & C.R. PRINGLE. 1976. Respiratory syncytial virus proteins. Virology **73:** 228–243.
49. GHILDYAL, R., J. MILLS, M. MURRAY, *et al.* 2002. Respiratory syncytial virus matrix protein associates with nucleocapsids in infected cells. J. Gen. Virol. **83:** 753–757.
50. EASTON, A.J. & P. CHAMBERS. 1997. Nucleotide sequence of the genes encoding the matrix and small hydrophobic proteins of pneumonia virus of mice. Virus Res. **8:** 27–33.
51. COLLINS, P.L. & G.W. WERTZ. 1985. The 1A protein gene of human respiratory syncytial virus: nucleotide sequence of the mRNA and a related polycistronic transcript. Virology **141:** 283–291.
52. TRIPP, R.A., L. JONES & L.J. ANDERSON. 2000. Respiratory syncytial virus G and/or SH glycoproteins modify CC and CXC chemokine mRNA expression in the BALB/c mouse. J. Virol. **74:** 6227–6229.
53. GARCIA-BARRENO, B., T. DELGADO & J.A. MELERO. 1994. Oligo(A) sequences of human respiratory syncytial virus G protein gene assessment of their genetic stability in frame shift mutants. J. Virol. **68:** 5460–5468.
54. MELERO, J.A., B. GARCIA-BARRENO, L. MARTINEZ, *et al.* 1997. Antigenic structure, evolution and immunobiology of human respiratory syncytial virus attachment (G) protein. J. Gen. Virol. **78:** 2411–2418.
55. ALVAREZ, R., H.M. LWAMBA, D.R. KAPCZYNSKI, *et al.* 2003. Nucleotide and predicted amino acid sequence-based analysis of the avian metapneumovirus type C cell attachment glycoprotein gene: phylogenetic analysis and molecular epidemiology of U.S. pneumoviruses. J. Clin. Microbiol. **41:** 1730–1735.
56. COLLINS, P.L., M.G. HILL, E. CAMARGO, *et al.* 1995. Production of infectious human respiratory syncytial virus from cloned cDNA confirms an essential role for the transcription elongation factor from the 5- proximal open reading

frame of the M2 mRNA in gene expression and provides a capability for vaccine development. Proc. Natl. Acad. Sci. USA **92:** 11563–11567.
57. GONZALEZ-REYES, L., M.B. RUIZ-ARGUELLO, B. GARCIA-BARRENO, *et al.* 2001. Cleavage of the human respiratory syncytial virus fusion protein at two distinct sites is required for activation of membrane fusion. Proc. Natl. Acad. Sci. USA **98:** 9859–9864.
58. ZIMMER, G., L. BUDZ & G. HERRLER. 2001. Proteolytic activation of respiratory syncytial virus fusion protein. Cleavage at two furin consensus sequences. J. Biol. Chem. **276:** 31642–31650.
59. ZIMMER, G., K.K. CONZELMANN & G. HERRLER. 2002. Cleavage at the furin consensus sequence RAR/KR (109) and presence of the intervening peptide of the respiratory syncytial virus fusion protein are dispensable for virus replication in cell culture. J. Virol. **76:** 9218–9224.
60. BEGONA RUIZ-ARGUELLO, M., L. GONZALEZ-REYES, L.J. CALDER, *et al.* 2002. Effect of proteolytic processing at two distinct sites on shape and aggregation of an anchorless fusion protein of human respiratory syncytial virus and fate of the intervening segment. Virology **298:** 317–326.
61. HARDY, R.W. & G.W. WERTZ, 1998. The product of the respiratory syncytial virus M2 gene ORF1 enhances read through of intergenic junctions during viral transcription. J. Virol. **72:** 520–526.
62. COLLINS, P.L., M.G. HILL, J. CRISTINA, *et al.* 1996. Transcription elongation factor of respiratory syncytial virus, a nonsegmented negative strand RNA virus. Proc. Natl. Acad. Sci. USA **93:** 81–85.
63. STEC, D.S., M.G. HILL III & P.L. COLLINS. 1991. Sequence analysis of the polymerase L gene of human respiratory syncytial virus and predicted phylogeny of nonsegmented negative-strand viruses. Virology **183:** 273–287.
64. POCH, O., L. SAUVAGET, M. DELARUE, *et al.* 1989. Identification of four conserved motifs among the RNA-dependent polymerase encoding proteins. EMBO J. **8:** 3867–3874.
65. POCH, O., B.M. BLUMBERG, L. BOUGUELERET, *et al.* 1990. Sequence comparison of five polymerases (L proteins) of unsegmented negative strand RNA viruses: theoretical assignment of functional domains. J. Gen. Virol. **71:** 1153–1162.
66. BAYON-AUBOYER, M.H., C. ARNAULD, D. TOQUIN, *et al.* 2000. Nucleotide sequences of the F, L and G protein genes of two non-A/non-B avian pneumoviruses (APV) reveal a novel APV subgroup. J. Gen. Virol. **81:** 2723–2733.
67. JUHAZ, K. & A.J. EASTON. 1994. Extensive sequence variation in the attachment (G) protein gene of avian pneumovirus: evidence for two distinct subgroups. J. Gen. Virol. **75:** 2873–2880.
68. BUYS, S.B. & J.H. DU PREEZ. 1980. A preliminary report on the isolation of a virus causing sinusitis in turkeys in South Africa and attempts to attenuate the virus. Turkeys **28:** 36–46.
69. BUYS, S.B., J.H. DU PREEZ & H.J. ELS. 1989a. The isolation and attenuation of a virus causing turkey rhinotracheitis in turkeys in South Africa. Onderstepoort J. Vet. Res. **56:** 87–98.
70. BUYS, S.B., J.H. DU PREEZ & H.J. ELS. 1989b. Swollen head syndrome in chickens: a preliminary on the isolation of a possible etiological agent. J. South Afr. Vet. Assoc. **60:** 221–222.
71. GIRAUD, P. & G. BENNEJEAN. 1986. Turkey rhinotracheitis in France: preliminary investigations on ciliostatic virus. Vet. Rec. **119:** 606–607.

72. McDougall, J.S. & J.K.A. Cook. 1986. Turkey rhinotracheitis: preliminary investigations. Vet. Rec. **118:** 206–207.
73. Naylor, C.J. & R.C. Jones 1993. Turkey rhinotracheitis: a review. Vet. Bull. **63:** 339–349.
74. Cook, J.K.A., M.M. Ellis, L. Ling, *et al.* 1993. Antigenic differentiation of strains of turkey rhinotracheitis virus using monoclonal antibodies. Avian Pathol. **22:** 257–273.
75. Tanaka, M., H. Takuma, N. Kokumai, *et al.* 1995. Turkey rhinotracheitis virus isolated from broiler chicken with swollen head syndrome in Japan. J. Vet. Med. Sci. **57:** 939–941.
76. Decanini, E.L., E.C. Miranda & F.X. Le Gros, 1991. Swollen head syndrome in heavy breeders in Mexico. Proceedings of the 40th Western Poultry Disease Conference, Acapulco, Mexico, pp. 158–159.
77. Weisman, Y., C. Strengel, R. Blumenkranz, *et al.* 1988. Turkey rhinotracheitis (TRT) in Turkey flocks in Israel-virus isolation and serological response. Proceedings of 37th Western Poultry Disease Conference, Davis, CA, pp. 67–69.
78. El Houadfi,, M., J. Hamam, J. Vanmarche, *et al.* 1991. Swollen head syndrome in broiler chicken in Morocco. Proceedings of 40th Western Poultry Disease Conference, Acapulco, Mexico, pp. 126–127.
79. Arns, C.W. & H.M. Hafez. 1992. Swollen head syndrome in poultry flocks in Brazil. Proceedings of the 41st Western Poultry Conference Sacramento, CA, pp 81–83.
80. Jones, R.C. 1996. Guest editorial. Avian pneumovirus infections: questions still unanswered. Avian Pathol. **25:** 639–648.
81. Senne, D.A., R.K. Edson, J.C. Pedersenet, *et al.* 1997. Avian pneumovirus update. Proceedings of American Veterinary Medical Association 134th Annual Congress, Reno, NV, USA, p. 190.
82. Cook, J.K.A., M.B. Huggins, S.J. Orbell, *et al.* 1999. Preliminary antigenic characterization of an avian Pneumovirus isolated from commercial turkeys in Colorado, USA. Avian Pathol. **28:** 607–617.
83. Stuart, J.C. 1989. *In* C. Nixey & T.C. Grey, Eds.: 217–224. Rhinotracheitis: Turkey rhinotracheitis (TRT) in Great Britain. Recent Advances in Turkey Science, Poultry Science Symposium, Series No. 21, London.
84. Kleven, S.H. 1997. Report of the Committee: transmissible diseases of poultry and other avian species. *In* Proceedings of the U.S. Animal Health Association 101st Annual Meeting. U.S. Animal Health Association, Washington, DC, pp. 486–491.
85. Shin, H.J., M.K. Njenga, B. McComb, *et al.* 2000. Avian pneumovirus RNA from wild and sentinel birds in US has genetic homology with APV isolates from domestic turkeys. J. Clin. Microbiol. **38:** 4282–4284.
86. Shin, H.J., K.V. Nagaraja, B. McComb, *et al.* 2002a. Isolation of avian pneumovirus from mallard ducks that is genetically similar to viruses isolated from neighboring commercial turkeys. Virus Res. **83:** 207–212.
87. Senne, D.A., J.C. Pedersen, R.K. Edson, *et al.* 1998. Avian pneumovirus in turkeys: have you seen it? Proc. West Poult. Dis. Conf. **47:** 67–68.
88. Toquin, D., M.H. Bayon-Auboyer, N. Eteradossi, *et al.* 1999a. Isolation of pneumovirus from a Muscovy duck. Vet. Rec. **145:** 680.
89. Gough, R.E., M.S. Collins, W.J. Cox, *et al.* 1988. Experimental infection of turkeys, chickens, ducks, geese, guinea fowl, pheasants, and pigeons with turkey rhinotracheitis virus. Vet. Rec. **123:** 58–59.

90. WELCHMAN, D.B., J.M. BRADBURY, D. CAVANAGH, *et al.* 2002. Infectious agents associated with respiratory disease in pheasants. Vet. Rec. **150:** 658–664.
91. PICAULT, J.-P., P. GIRAUD, M. GUITTET, *et al.* 1987. Isolation of a TRTV-like virus from chickens with swollen head syndrome. Vet. Rec. **121:** 135.
92. CADMAN, H.F., P.J. KELLY, R. ZHOU, *et al.* 1994. A serosurvey using enzyme-linked immunosorbent assay for antibodies against poultry pathogens in Ostriches (*Struthio camelus*) from Zimbabwe. Avian Dis. **38:** 621–625.
93. JIRJIS, F.E., S.L. NOLL, D.A. HALVORSON, *et al.* 2000. Avian pneumovirus infection in Minnesota turkeys: experimental reproduction of the disease. Avian Dis. **44:** 222–226.
94. BASTIEN, N., L. LIU, D. WARD, *et al.* 2004. Genetic variability of the G glycoprotein gene of human metapneumovirus. J. Clin. Microbiol. **42:** 3532–3537.
95. BELL, I.G. & D.J. ALEXANDER. 1990. Failure to detect antibody to turkey rhinotracheitis virus in Australia poultry flocks. Aust. Vet. J. **67:** 232–233.
96. PERET, T.C., G. BOIVIN, Y. LI, *et al.* 2002. Characterization of human metapneumoviruses isolated from patients in North America. J. Infect. Dis. **185:** 1660–1663.
97. VAN DEN HOOGEN, B.G., A.D. OSTERHAUS & A.M. FOUCHIER. 2004. Clinical impact and diagnosis of human metapneumovirus infection. Pediatr. Infect. Dis. J. **23:** S25–S32.
98. BIACCHESI, S., M.H. SKIADOPOULOS, G. BOIVIN, *et al.* 2003. Genetic diversity between human metapneumovirus subgroups. Virology **315:** 1–9.
99. PERET, T.C., Y. ABED, L.J. ANDERSON, *et al.* 2004. Sequence polymorphism of the predicted human metapneumovirus G glycoprotein. J. Gen. Virol. **85:** 679–686.
100. HUCK, B., G. SCHARF, D. NEUMANN-HAEFELIN, *et al.* 2006. Novel human metapneumovirus sublineage. Emerg. Infect. Dis. **12:** 147–150.
101. BOIVIN, G., Y. ABED, G. PELLETIER, *et al.* 2002. Virological features and clinical manifestations associated with human metapneumovirus: a new paramyxovirus responsible for acute respiratory-tract infections in all age groups. J. Infect. Dis. **186:** 1330–1334.
102. CUEVAS, L.E., A.M. NASSER, W. DOVE, *et al.* 2003. Human metapneumovirus and respiratory syncytial virus, Brazil. Emerg. Infect. Dis. **9:** 1626–1628.
103. EBIHARA, T., R. RIKA ENDO, H. KIKUTA, *et al.* 2004. Human metapneumovirus infection in Japanese children. J. Clin. Microbiol. **42:** 126–132.
104. LUDEWICK, P.H., Y. ABED, N. VAN NIEKERK, *et al.* 2005. Human metapnemovirus genetic variability, South Africa. Emerg. Infect. Dis. **11:** 1074–1078.
105. MAGGI, F., M. PIFFERI, M. VATTERONI, *et al.* 2003. Human metapneumovirus associated with respiratory tract infections in a 3-year study of nasal swabs from infants in Italy. J Clin Microbiol. **41:** 2987–2991.
106. PELLETIER, G., P. DÉRY, Y. ABED, *et al.* 2002. Respiratory tract reinfections by the new human metapneumovirus in an immunocompromised child. Emerg. Infect. Dis. **8:** 976–978.
107. STOCKON, J., I. STEPHENSON, D. FLEMING, *et al.* 2002. Human metapmneumovirus as a cause of community-acquired respiratory illness. Emerg. Infect. Dis. **8:** 897–901.
108. BANERJEE, S., P. BHARAJ P, W. SULLENDER, *et al.* 2007. Human metapneumovirus infections among children with acute respiratory infections seen in a large referral hospital in India. J. Clin. Virol. **38:** 70–72.

109. RAO, B.L., S.S. GANDHE, S.D. PAWAR, *et al.* 2004. First detection of human metapneumovirus in children with acute respiratory infection in India: a preliminary report. J. Clin. Microbiol. **42:** 5961–5962.
110. BOIVIN, G., G. DE SERRES, S. COTÉ, *et al.* 2003. Human metapneumovirus infections in hospitalized children. Emerg. Infect. Dis. **9:** 634–640.
111. VAN DEN HOOGEN, B.G., G.J.J. VAN DOORNUM, J.C. FOCKENS, *et al.* 2003. Prevalence and clinical symptoms of human metapneumovirus infection in hospitalized patients. J. Infect. Dis. **188:** 1571–1577.
112. FALSEY, A.R., D. ERDMAN, L.J. ANDERSON, *et al.* 2003. Human metapneumovirus infections in young and elderly adults. J. Infect. Dis. **187:** 785–790.
113. PEIRIS, J.S., W.H. TANG, K.H. CHAN, *et al.* 2003. Children with respiratory disease associated with metapneumovirus in Hong Kong. Emerg. Infect. Dis. **9:** 628–633.
114. WILLIAMS, J.V., P.A. HARRIS, S.J. TOLLEFSON, *et al.* 2004. Human metapneumovirus and lower respiratory tract disease in otherwise healthy infants and children. NEJM **350:** 443–450.
115. VIAZOV, S., F. RATJEN, R. SCHEIDHAUER, *et al.* 2003. High prevalence of human metapneumovirus infection in young children and genetic heterogeneity of the viral isolates. J. Clin. Microbiol. **41:** 3043–3045.
116. CANE, P.A., B.G. VAN DEN HOOGEN, S. CHAKRABARTI, *et al.* 2003. Human metapneumovirus in a haemopoietic stem cell transplant recipient with fatal lower respiratory tract disease. Bone Marrow Transplant. **31:** 309–310.
117. CROWE, J.E.J. & J.V. WILLIAMS. 2003. Immunology of viral respiratory tract infection in infancy. Paediatr. Respir. Rev. **4:** 112–119.
118. JARTTI, T., B.G. VAN DEN HOOGAN, R.P. GAROFALO, *et al.* 2002. Metapneumovirus and acute wheezing in children. Lancet **360:** 1393–1394.
119. GREENSILL, J., P.S. MCNAMARA, W. DOVE, *et al.* 2003. Human metapneumovirus in severe respiratory syncytial virus bronchiolitis. Infect. Dis. **9:** 372–375.
120. COTE, S., Y. ABED & G. BOIVIN. 2003. Comparative evaluation of real-time PCR assays for detection of the human metapneumovirus. J. Clin. Microbiol. **41:** 3631–3635.
121. PRINS, J.M. & K.C. WOLTHERS. 2004. Human metapneumovirus: a new pathogen in children and adults. Netherlands J. Med. **62:** 177–179.

Signal Transduction in SARS-CoV-Infected Cells

T. MIZUTANI

Department of Virology 1, National Institute of Infectious Diseases, Musashimurayama, Tokyo, Japan

ABSTRACT: Severe acute respiratory syndrome (SARS) is a newly found infectious disease that is caused by a novel human coronavirus, SARS coronavirus (SARS-CoV). Because the mortality rate of SARS patients is very high, understanding the pathological mechanisms of SARS not only *in vivo* but *in vitro* is important for the prevention of SARS. Activation of signaling pathways caused by SARS-CoV infection leads to the phosphorylation and activation of downstream molecules. Two conflicting cellular programs, apoptosis to eliminate virus-infected cells and survival to delay apoptosis by producing antiviral cytokines, occur in SARS patients. Recent studies regarding SARS and SARS-CoV have clarified that activation of mitogen-activated protein kinases (MAPKs) plays important roles in upregulation of cytokine expression and apoptosis both *in vitro* and *in vivo*. Both Akt and p38 MAPK are keys for determination of cell survival or death in SARS-CoV-infected cells *in vitro*. Agents being developed to target these signaling cascades may be important for the design of anti-SARS-CoV drugs. This review highlights recent progress regarding SARS-CoV biology, especially signal transduction in SARS-CoV-infected cells.

KEYWORDS: SARS; apoptosis; MAPK

INTRODUCTION

Severe acute respiratory syndrome (SARS) is a newly found infectious disease that is caused by a novel coronavirus, SARS coronavirus (SARS-CoV).[26,34] In late 2002, SARS spread from China to more than 30 countries, causing severe outbreaks of atypical pneumonia. Although the mechanism of SARS pathogenesis *in vivo* may involves both the effects of viral replication in the target cells and immune responses to viral antigens, there is a lack of molecular pathological data, including data regarding the signaling pathways of SARS-CoV infection. As viral virulence and the mortality rate of SARS

Address for correspondence: T. Mizutani, Department of Virology 1, National Institute of Infectious Diseases, Gakuen 4-7-1, Musashimurayama, Tokyo 208-0011, Japan. Voice: +81-42-561-0771; fax: +81-42-564-4881.
tmizutan@nih.go.jp

Ann. N.Y. Acad. Sci. 1102: 86–95 (2007).
doi: 10.1196/annals.1408.006

patients are very high, understanding the pathological mechanisms of SARS is important for the prevention of SARS.

Generally, both proapoptotic and prosurvival signaling pathways are activated during viral replication. Many pro- and antiapoptotic proteins are involved in these pathways in cells. Several reports indicated that activation of physiological intracellular signaling cascades caused by SARS-CoV infection leads to the phosphorylation and activation of downstream molecules. For example, mitogen-activated protein kinases (MAPKs) are well-known signal transducers that respond to extracellular stimulation by cytokines, growth factors, viral infection, and stress, and in turn regulate cell differentiation, proliferation, survival, and apoptosis.[15,23,40] Recent studies regarding SARS and SARS-CoV have clarified that activation of MAPKs plays important roles in upregulation of cytokine expression and apoptosis both *in vitro* and *in vivo*. This review highlights progress regarding SARS-CoV biology, especially signal transduction in SARS-CoV-infected cells (FIG. 1).

p38 MAPK

Apoptosis, which is fundamentally different from necrosis, is an active and physiological type of cell death, which can be induced by viral infection, viral replication, and production of viral proteins. The monkey kidney cell line, Vero E6, is widely used in SARS-CoV research due to its high sensitivity to infection with the virus. Several studies have shown that SARS-CoV infection of Vero E6 cells induces apoptosis, detected by DNA fragmentation and caspase activation.[28,42] p38 MAPK is strongly activated by stress and inflammatory cytokines. Mouse hepatitis virus (MHV), a prototype coronavirus, was able to induce activation of p38 MAPK into virus infected cells.[1] p38 MAPK activation in CD14 monocytes was also observed in SARS patients.[24] Phosphorylation of p38 MAPK was significantly upregulated at 18 h postinfection (h p.i.) in SARS-CoV-infected Vero E6 cells.[28] Cytopathic effects of SARS-CoV-infection were partially prevented by adding the p38 MAPK-specific inhibitor, SB203580, to the cells. Although p38 MAPK can promote both cell death and survival,[10] the results of inhibitor studies indicated that the p38 MAPK signaling pathway may be involved mainly in cell death in SARS-CoV-infected Vero E6 cells. Several downstream targets of p38 MAPK were phosphorylated in virus-infected cells, and SB203580 effectively inhibited phosphorylation of these proteins in SARS-CoV-infected cells. MAPK-activated protein kinase (MAPKAPK)-2, which is activated in response to stress and growth factors,[12,13] was phosphorylated in virus-infected cells. Hsp-27, which is a substrate of MAPKAPK-2 and is known to show antiapoptotic activity by inhibiting apoptosome formation,[14] survival factors,[20] cAMP response element-binding protein (CREB), and activation of transcription factor (ATF)-1, was also phosphorylated in virus-infected cells. Phosphorylation of Hsp-27, CREB, and ATF-1 may induce an antiapoptotic environment in

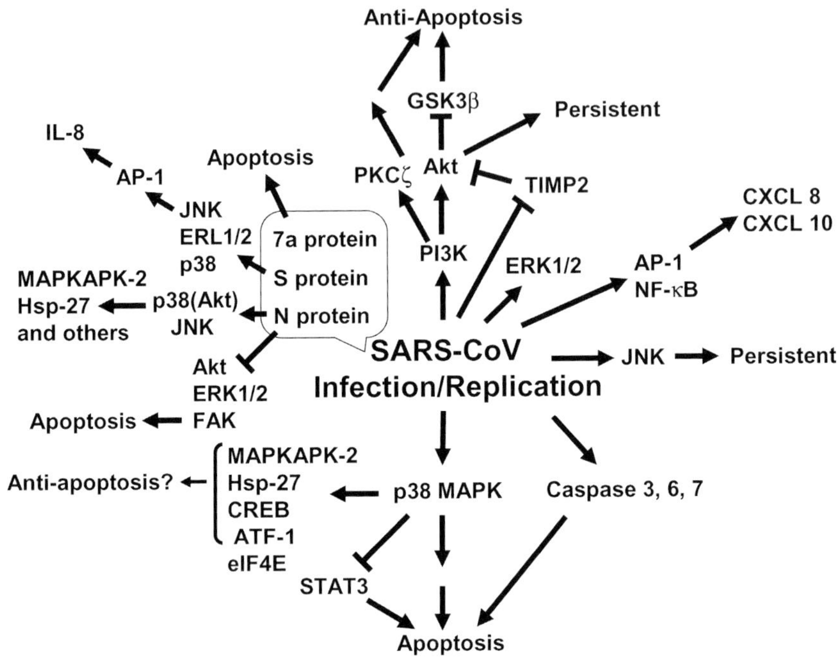

FIGURE 1. Schematic representation of signal transduction on SARS-CoV infection.

SARS-CoV-infected cells. Nucleocapsid (N) protein was able to induce phosphorylation of Hsp-27 and CREB in transfected cells as described below.[36] The translation initiation factor, eukaryotic initiation factor 4E (eIF4E), is known to enhance translation rates of cap-containing mRNAs.[16] The phosphorylation of eIF4E was utilized to promote virus-specific protein synthesis in the case of MHV.[1] Although the levels of phosphorylated eIF4E were increased by SARS-CoV infection, the activated eIF4E was not advantageous for viral protein synthesis as demonstrated by the similar kinetics of viral protein accumulation in infected Vero E6 cells in the presence and absence of SB 203580. There may be other substrates of p38 MAPK that are inducible on apoptosis of Vero E6 cells caused by SARS-CoV infection. As each downstream target molecule of p38 MAPK has a role in the induction of cell death or survival in response to SARS-CoV infection, determination of the apoptosis-inducible target molecules of p38 MAPK is important for the development of anti-SARS-CoV drugs.

ERK1/2

Extracellular signal-regulated kinase (ERK) 1/2 was phosphorylated in SARS-CoV-infected Vero E6 cells,[27] whereas ERK1/2 was downregulated in N protein-expressing COS-1 cells as described below.[36]

JNK

c-Jun N-terminal kinase (JNK) was phosphorylated in SARS-CoV-infected Vero E6 cells.[27] A recent study indicated that persistent infection was established after most of the SARS-CoV-infected Vero E6 cells had died by apoptosis.[30] SP600125, an inhibitor of JNK, and LY294002, an inhibitor of phosphatidylinositol 3-kinase (PI3K), inhibited the establishment of persistence, whereas PD98059, an inhibitor of MEK1/2, and SB203580, an inhibitor of p38 MAPK, did not. Thus, two signaling pathways of JNK and PI3K are important for the establishment of persistence in Vero E6 cells.

STAT3

In Vero E6 cells, signal transducer and activator of transcription (STAT) 3 is constitutively phosphorylated at tyrosine (Tyr)-705 and is slightly phosphorylated at serine (Ser)-727.[27] However, SARS-CoV replication induced dephosphorylation of STAT3 at Tyr-705. As STAT3 is a major transcription factor activated in response to cytokines, such as interleukin-6 (IL-6) and IL-10, inhibition of STAT3 signaling by dominant negative and antisense oligonucleotides against STAT3 resulted in decreases in cell viability and apoptosis.[17,31] Thus, STAT3 is thought to act as an antiapoptotic transcription factor. Although inhibitors of MEK and JNK had no effect on the phosphorylation status of STAT3 in virus-infected cells, two inhibitors of p38 MAPK (SB203580 and SB202190) partially inhibited dephosphorylation of STAT3 at Tyr-705, suggesting that the p38 MAPK signaling pathway is upstream of Tyr-705 dephosphorylation of STAT3 in Vero E6 cells. The level of STAT3 phosphorylation at Ser-727 was increased in virus-infected cells. Although the effect of Ser-727 phosphorylation on the function of STAT3 remains unresolved, it was reported that phosphorylation of Ser-727 of STAT3 negatively modulated its Tyr phosphorylation.[6] Interestingly, Tyr-705 dephosphorylation and Ser-727 phosphorylation showed almost the same timing in SARS-CoV-infected cells. STAT3 phosphorylated at Tyr-705 was localized mainly in the nucleus in mock-infected cells, whereas STAT3 disappeared from the nucleus in virus-infected cells. As STAT3 acts as an activator of transcription in the nucleus, STAT3 may lack its transcriptional activity in SARS-CoV-infected Vero E6 cells, and thus result in a decrease in antiapoptotic activity in the cells.

As other STAT signal transduction pathways are involved in SARS-CoV infection, there have been many reports that treatment with interferons (IFNs) can inhibit viral replication *in vivo* and *in vitro*. However, there have been no detailed studies of IFN signal transduction in SARS-CoV-infected cells. IFN-alpha receptor recognizes STAT1 and STAT2 via Jak1 and Tyk2, whereas the IFN-gamma receptor recognizes STAT1 via Jak1 and Jak2. Signal-transducing adaptor molecule 1 (STAM1) was upregulated in SARS-CoV-infected Vero E6 cells.[25] As STAM1 is known to associate with Jak2 and 3 via the immunoreceptor Tyr-based activation motif, it may play an important role in signal transduction of cytokine receptors by SARS-CoV infection.

AKT

As mentioned above, the PI3K signaling pathway (including Akt) plays important roles in establishing persistent infection by SARS-CoV. Akt is phosphorylated at both Ser-473 and threonine (Thr)-308 residues via a PI3K-dependent mechanism on stimulation by growth factors, insulin, and hormones.[39] One of the most important functions of activated Akt in cells is the prevention of apoptosis. Akt has many downstream targets and it induces cell survival via phosphorylation of the forkhead transcription factor (FKHR) family, glycogen synthase kinase 3ß (GSK-3ß), caspase-9, and Bad.[2,8,9] In SARS-CoV-infected confluent Vero E6 cells, Ser-473 of Akt was phosphorylated at 8 h.p.i. and maximal phosphorylation was observed at 18 h.p.i., whereas phosphorylation of Thr-308 was not observed. As Akt is thought to show its full activity when phosphorylated at both Ser-473 and Thr-308, the total activity of Akt may be low in Vero E6 cells. Therefore, the phosphorylated Akt in SARS-CoV-infected cells cannot prevent apoptosis.

Interestingly, tissue inhibitor of metalloproteinase 2 (TIMP2), which activates Ras and leads to Ras/PI3K complex formation, was shown to be downregulated in virus-infected Vero E6 cells at 12 h.p.i. by microarray analysis.[25] The acute Ser dephosphorylation of Akt at 24 h.p.i. after tentative phosphorylation in virus-infected cells may be due to downregulation of TIMP2.

PKC

Protein kinase C (PKC) is one of the major cellular mediators of biological functions. The PKC superfamily is classified as subsuperfamilies based on activation profiles: conventional PKC (cPKC α, βI, βII, γ novel PKC (nPKC δ, ϵ, η, θ), atypical PKC (aPKC ζ, ι/λ) PKCμ PKD and PKCν. PKCζ, which was discovered as a unique PKC isotype,[32] is thought to be one of the most important PKC, because PI3K-dependent activation of PKCζ mediates B-cell survival by nerve growth factor.[22] Akt has been reported to interact with PKCζ.[21] PKCζ was phosphorylated in SARS-CoV-infected Vero E6 cells,[27] suggesting that this molecule is activated as an antiapoptotic response to SARS-CoV infection.

TRANSCRIPTION FACTORS

Human intestinal epithelial Caco-2 cells, which are highly permissive to SARS-CoV, have been used for analysis of cellular gene expression by microarray analysis. The proinflammatory chemokines, interleukin 8 (CXCL8), and interferon-γ-inducible protein 10 (CXCL10), were upregulated in SARS-CoV-infected Caco-2 cells.[7] These two chemokines are regulated by AP-1 and NF-κB, which were also activated by viral infection. CXCL10 levels were significantly elevated in the blood of SARS patients[38,41] and in macrophages *in vitro*.[5] As described above and below, CREB was phosphorylated in virus-infected Vero E6 cells and N protein-expressing COS-1 cells.

APOPTOSIS-INDUCIBLE VIRAL GENES

The genome of SARS-CoV is approximately 30 kb in length and contains 14 potential open-reading frames (ORFs), including nine viral proteins with no homologues in other coronaviruses. The N protein, which is a 423 amino acid predicted phospho-protein of 46 kDa with a short Ser-rich stretch and a putative bipartite nuclear localization signal, is known to show little homology with molecules in other coronaviruses. The N protein has been shown to undergo self-association through the C-terminal 209 amino acid region[35] and was able to induce apoptosis of COS-1 cells in the absence of growth factors.[36] Both JNK and p38 MAPK were upregulated in the transfected cells, whereas ERK and Akt were downregulated. Activated p38 MAPK by N protein induced actin reorganization into cells in the absence of growth factors. The downstream targets of p38 MAPK, MAPKAPK-2, and Hsp-27, were phosphorylated in the cells as well as in the virus-infected Vero E6 cells. Downregulation of focal adhesion kinase (FAK) activity and fibronectin expression was observed due to N protein expression, supporting the hypothesis that apoptosis induced by N protein is caused by interference with the integrin signaling pathway, and that serum factors are able to inhibit apoptosis by maintaining the activity of FAK through alternative pathways. Caspase-3 and -7 were activated only in the absence of growth factors in N-expressing cells. However, SB203580, an inhibitor of p38 MAPK, failed to inhibit caspase-3 activation. Thus, there are two independent functional properties of N protein, p38 MAPK-dependent actin reorganization, and caspase activation. Another group showed that the levels of transcription factors, c-Fos, ATF2, CREB, and FosB, are increased by expression of N in Vero and Huh-7 cells.[18] Spike (S) protein of SARS-CoV also induced activation of AP-1 and IL-8 promoter, possibly via activation of MAPKs, in A549 and HFL-1 cells.[4] As high serum levels of IL-8 were observed in patients in the acute stage of SARS,[19] activation of the IL-8 promoter via AP-1 induced by S protein may explain these clinical observations. The binding of S protein to a viral receptor, angiotensin-converting enzyme (ACE)-2, may be a trigger for activation of these MAPKs.

Two unique proteins of SARS-CoV, 3a (U274, X1) and 7a (U122, X4), have been studied. The 3a protein is located mainly in the Golgi apparatus and contains three transmembrane regions,[43] whereas 7a encodes a protein of 122 amino acids containing a probable cleaved signal sequence and a C-terminal transmembrane helix. As the C-terminal tail also contains a typical endoplasmic reticulum (ER) retrieval motif, 7a is localized to the perinuclear region in both SARS-CoV-infected and -transfected cells.[11] The overexpression of 7a, but not of 3a, induced apoptosis via a caspase-dependent pathway.[37]

CONCLUSIONS

In SARS patients, two conflicting cellular programs occur: "apoptosis" to eliminate virus-infected cells and "survival" to delay apoptosis by producing antiviral cytokines. The control mechanisms balancing cell survival against cell death *in vitro* are important for understanding the pathology *in vivo*. As mentioned above, activation of p38 MAPK is involved in the regulation of IL-8 *in vivo*. Both Akt and p38 MAPK are keys for determination of cell survival or death in SARS-CoV-infected cells *in vitro*. Activation of the p38 MAPK signaling pathway and dephosphorylation of STAT3 via p38 MAPK induced by SARS-CoV infection have partially proapoptotic roles in Vero E6 cells. On the other hand, the levels of Akt, which inactivates proapoptotic pathways, are too low to block apoptosis signaling pathways. As Akt activation is necessary to establish persistently infected cells that escape from apoptotic cell death on viral infection, Akt plays an important role in delaying apoptosis in SARS-CoV-infected cells. Agents being developed to target p38 MAPK and its downstream molecules (for downregulation) and Akt (for upregulation) may be important for the design of anti-SARS-CoV drugs.

ADDENDUM

In recent papers, Surjit *et al*. reported that phosphorylated N protein shuttled from nucleus to cytoplasm by binding to 14-3-3 (J. Virol. 2005. 79: 11476–11486) and N protein blocked S phase progression (J. Biol. Chem. 2006. 281: 10669–10681). Yuan *et al*. reported that 7a protein blocked cell cycle progression at G0/G1 phase (Virology 2006. 346. 74–85). Inhibition of cell proliferation was caused by dephosphorylation of Akt (by us; FEMS Immunol. Med. Microbiol. 2006. 46: 236–243). Okabayashi *et al*. reported that production and response of IFN were not suppressed by SARS-CoV-infection (J. Med. Virol. 2006. 78: 417–424). We reported regulation 90 kDa ribosomal S6 kinase phosphorylation (FEBS Lett. 2006. 580: 1417–1424), importance of Bcl-2 and Bcl-xL for the establishment of persistent infection (Biochem.Biophys.Res.Commun. 2006. 347: 261–265), and enhancement of cytotoxity against Vero E6 cells persistently infected with SARS-CoV by *Mycoplasma fermentans* (Arch. Virol. In press).

ACKNOWLEDGMENTS

I thank Drs. S. Fukushi, M. Saijo, M. Ogata, K. Sakai, I. Kurane, and S. Morikawa (National Institute of Infectious Diseases, Japan) for useful suggestions. We acknowledge funding from a Grant-in-Aid for Scientific Research from the Ministry of Education, Science, Sports and Culture of Japan, a grant-in-aid from the Ministry of Health, Labor, and Welfare of Japan, and the Japan, Health Science Foundation, Tokyo, Japan.

REFERENCES

1. BANERJEE, S., K. NARAYANAN, T. MIZUTANI & S. MAKINO. 2002. Murine coronavirus replication-induced p38 mitogen-activated protein kinase activation promotes interleukin-6 production and virus replication in cultured cells. J. Virol. **76:** 5937–5948.
2. CARDONE, M.H., N. ROY, H.R. STENNICKE, *et al.* 1998. Regulation of cell death protease caspase-9 by phosphorylation. Science **282:** 1318–1321.
3. CHANG, L. & M. KARIN. 2001. Mammalian MAP kinase signalling cascades. Nature **410:** 37–40.
4. CHANG, Y.J., C.Y. LIU, B.L. CHIANG, *et al.* 2004. Induction of IL-8 release in lung cells via activator protein-1 by recombinant baculovirus displaying severe acute respiratory syndrome-coronavirus spike proteins: identification of two functional regions. J. Immunol. **173:** 7602–7614.
5. CHEUNG, C.Y., L.L. POON, L.H. NG, *et al.* 2005. Cytokine responses in severe acute respiratory syndrome coronavirus-infected macrophages *in vitro*: possible relevance to pathogenesis. J. Virol. **79:** 7819–7826.
6. CHUNG, J., E. UCHIDA, T.C. GRAMMER & J. BLENIS. 1997. STAT3 serine phosphorylation by ERK-dependent and -independent pathways negatively modulates its tyrosine phosphorylation. Mol. Cell. Biol. **17:** 6508–6516.
7. CINATL, J. JR., M. MICHAELIS, B. MORGENSTERN & H.W. DOERR. 2005. High-dose hydrocortisone reduces expression of the pro-inflammatory chemokines CXCL8 and CXCL10 in SARS coronavirus-infected intestinal cells. Int. J. Mol. Med. **15:** 323–327.
8. CROSS, D.A., D.R. ALESSI, P. COHEN, *et al.* 1995. Inhibition of glycogen synthase kinase-3 by insulin mediated by protein kinase B. Nature **378:** 785–789.
9. DATTA, S.R., H. DUDEK, X. TAO, *et al.* 1997. Akt phosphorylation of BAD couples survival signals to the cell-intrinsic death machinery. Cell **91:** 231–241.
10. DENT, P., A. YACOUB, P.B. FISHER, *et al.* 2002. MAPK pathways in radiation responses. Oncogene **22:** 5885–5896.
11. FIELDING, B.C., Y.J. TAN, S. SHUO, *et al.* 2004. Characterization of a unique group-specific protein (U122) of the severe acute respiratory syndrome coronavirus. J. Virol. **78:** 7311–7318.
12. FOLTZ, I.N., J.C. LEE, P.R. YOUNG & J.W. SCHRADER. 1997. Hemopoietic growth factors with the exception of interleukin-4 activate the p38 mitogen-activated protein kinase pathway. J. Biol. Chem. **272:** 3296–3301.
13. FRESHNEY, N.W., L. RAWLINSON, F. GUESDON, *et al.* 1994. Interleukin-1 activates a novel protein kinase cascade that results in the phosphorylation of Hsp27. Cell **78:** 1039–1049.

14. GARRIDO, C., E. SCHMITT, C. CANDE, *et al.* 2003. HSP27 and HSP70: potentially oncogenic apoptosis inhibitors. Cell Cycle **2:** 579–584.
15. GARRINGTON, T.P. & G.L. JOHNSON. 1999. Organization and regulation of mitogen-activated protein kinase signaling pathways. Curr. Opin. Cell. Biol. **11:** 211–218.
16. GINGRAS, A.C., B. RAUGHT & N. SONENBERG. 1999. eIF4 initiation factors: effectors of mRNA recruitment to ribosomes and regulators of translation. Annu. Rev. Biochem. **68:** 913–963.
17. GRANDIS, J.R., S.D. DRENNING, Q. ZENG, *et al.* 2000. Constitutive activation of Stat3 signaling abrogates apoptosis in squamous cell carcinogenesis *in vivo*. Proc. Natl. Acad. Sci. USA **97:** 4227–4232.
18. HE, R., A. LEESON, A. ANDONOV, *et al.* 2003. Activation of AP-1 signal transduction pathway by SARS coronavirus nucleocapsid protein. Biochem. Biophys. Res. Commun. **311:** 870–876.
19. HUANG, K.J., I.J. SU, M. THERON, *et al.* 2005. An interferon-gamma-related cytokine storm in SARS patients. J. Med. Virol. **75:** 185–194.
20. JEAN, D., M. HARBISON, D.J. MCCONKEY, *et al.* 1998. CREB and its associated proteins act as survival factors for human melanoma cells. J. Biol. Chem. **273:** 24884–24890.
21. KONISHI, H., T. SHINOMURA, S. KURODA, *et al.* 1994. Molecular cloning of rat RAC protein kinase alpha and beta and their association with protein kinase C zeta. Biochem. Biophys. Res. Commun. **205:** 817–825.
22. KRONFELD, I., G. KAZIMIRSKY, E.W. GELFAND & C. BRODIE. 2002. NGF rescues human B lymphocytes from anti-IgM induced apoptosis by activation of PKC. Eur. J. Immunol. **32:** 136–143.
23. KYRIAKIS, J.M. & J. AVRUCH. 2001. Mammalian mitogen-activated protein kinase signal transduction pathways activated by stress and inflammation. Physiol. Rev. **81:** 807–869.
24. LEE, C.H., R.F. CHEN, J.W. LIU, *et al.* 2004. Altered p38 mitogen-activated protein kinase expression in different leukocytes with increment of immunosuppressive mediators in patients with severe acute respiratory syndrome. J. Immunol. **172:** 7841–7847.
25. LEONG, W.F., H.C. TAN, E.E. OOI, *et al.* 2005. Microarray and real-time RT-PCR analyses of differential human gene expression patterns induced by severe acute respiratory syndrome (SARS) coronavirus infection of Vero cells. Microbes Infect. **7:** 248–259.
26. MARRA, M.A., S.J. JONES, C.R. ASTELL, *et al.* 2003. The genome sequence of the SARS-associated coronavirus. Science **300:** 1399–1404.
27. MIZUTANI, T., S. FUKUSHI, M. MURAKAMI, *et al.* 2004. Tyrosine dephosphorylation of STAT3 in SARS coronavirus-infected Vero E6 cells. FEBS Lett. **577:** 187–192.
28. MIZUTANI, T., S. FUKUSHI, M. SAIJO, *et al.* 2004. Phosphorylation of p38 MAPK and its downstream targets in SARS coronavirus-infected cells. Biochem. Biophys. Res. Commun. **319:** 1228–1234.
29. MIZUTANI, T., S. FUKUSHI, M. SAIJO, *et al.* 2004. Importance of Akt signaling pathway for apoptosis in SARS-CoV-infected Vero E6 cells. Virology **327:** 169–174.
30. MIZUTANI, T., S. FUKUSHI, M. SAIJO, *et al.* 2005. JNK and PI3K/Akt signaling pathways are required for establishing persistent SARS-CoV-infection in Vero E6 cells. Biochim. Biophys. Acta **1741:** 4–10.

31. MORA, L.B., R. BUETTNER, J. SEIGNE, *et al.* 2002. Constitutive activation of Stat3 in human prostate tumors and cell lines: direct inhibition of Stat3 signaling induces apoptosis of prostate cancer cells. Cancer Res. **62:** 6659–6666.
32. ONO, Y., T. FUJII, K. OGITA, *et al.* 1989. Protein kinase C- subspecies from brain: its structure, expression and properties. Proc. Natl. Acad. Sci. USA **86:** 3099–3103.
33. RAJAN, P. & R.D. MCKAY. 1998. Multiple routes to astrocytic differentiation in the CNS. J. Neurosci. **18:** 3620–3629.
34. ROTA, P.A., M.S. OBERSTE, S.S. MONROE, *et al.* 2003. Characterization of a novel coronavirus associated with severe acute respiratory syndrome. Science **300:** 1394–1399.
35. SURJIT, M., B. LIU, P. KUMAR, *et al.* 2004. The nucleocapsid protein of the SARS coronavirus is capable of self-association through a C-terminal 209 amino acid interaction domain. Biochem. Biophys. Res. Commun. **317:** 1030–1036.
36. SURJIT, M., B. LIU, S. JAMEEL, *et al.* 2004. The SARS coronavirus nucleocapsid (N) protein induces actin reorganization and apoptosis in COS-1 cells. Biochem. J. **383:** 13–18.
37. TAN, Y.J., B.C. FIELDING, P.Y. GOH, *et al.* 2004. Overexpression of 7a, a protein specifically encoded by the severe acute respiratory syndrome coronavirus, induces apoptosis via a caspase-dependent pathway. J. Virol. **78:** 14043–14047.
38. WARD, S.E., M.R. LOUTFY, L.M. BLATT, *et al.* 2005. Dynamic changes in clinical features and cytokine/chemokine responses in SARS patients treated with interferon alfacon-1 plus corticosteroids. Antivir. Ther. **10:** 263–275.
39. WELCH, H., A. EGUINOA, L.R. STEPHENS & P.T. HAWKINS. 1998. Protein kinase B and RAC are activated in parallel within a phosphatidylinositide 3OH-kinase-controlled signaling pathway. J. Biol. Chem. **273:** 11248–11256.
40. WHITMARSH, A.J. & R.J. DAVIS. 2000. A central control for cell growth. Nature **403:** 255–256.
41. WONG, C.K., C.W. LAM, A.K. WU, *et al.* 2004. Plasma inflammatory cytokines and chemokines in severe acute respiratory syndrome. Clin. Exp. Immunol. **136:** 95–103.
42. YAN, H., G. XIAO, J. ZHANG, *et al.* 2004. SARS coronavirus induces apoptosis in Vero E6 cells. J. Med. Virol. **73:** 323–331.
43. YUAN, X., J. LI, Y. SHAN, *et al.* 2005. Subcellular localization and membrane association of SARS-CoV 3a protein. Virus Res. **109:** 191–202.

The Molecular and Cellular Aspects of Arthritis Due to Alphavirus Infections

Lesson Learned from Ross River Virus

NESTOR E. RULLI,[a] JULIAN MELTON,[b] ANJA WILMES,[a] GARY EWART,[b] AND SURESH MAHALINGAM[a]

[a]*Centre for Virus Research, School of Health Sciences, University of Canberra, Canberra, ACT 2601, Australia*

[b]*Biotron Limited, LPO Box 8315, Australian National University, Canberra, ACT 2601, Australia*

ABSTRACT: **Alphaviruses such as the Sindbis-group viruses, Scandinavian Ockelbo virus, the African Asian chikungunya virus, the African O'nyong-nyong virus, the South American Mayaro virus, and the Australasian Barmah Forest and Ross River viruses, are commonly associated with outbreaks of acute and persistent arthritis and arthralgia in humans. The mechanisms by which these viruses cause arthritis/arthralgia are poorly understood. This chapter summarizes our current understanding of viral arthritides using our newly developed mouse model of Ross River virus-induced joint and muscle inflammation.**

KEYWORDS: **alphavirus; Ross River virus; mouse model of disease; viral arthritis; macrophages**

INTRODUCTION

There are 26 known members of alphaviruses and all of them are arboviruses.[1] Alphaviruses have a wide host range: antibodies have been found in marsupials, rodents, reptiles, fish, birds, humans, horses, and domestic pets. However, humans are almost exclusively incidental (dead end) hosts and are only required for maintenance of alphaviruses in exceptional circumstances (TABLE 1).[2]

The enzootic (maintenance) cycle of most alphaviruses seems to involve only a single vector and single host species in one geographical area. The level of virus circulating in the blood of infected hosts is sufficient to infect the vector species, which in turn infects vertebrate hosts efficiently enough to ensure continuous virus propagation throughout the transmission season. In the natural

Address for correspondence: Suresh Mahalingam, School of Health Sciences, University of Canberra, Canberra, ACT 2601, Australia. Voice: 61-02-62012368; fax: 61-02-62015727. suresh.mahalingam@canberra.edu.au

Ann. N.Y. Acad. Sci. 1102: 96–108 (2007).
doi: 10.1196/annals.1408.007

TABLE 1. Alphaviruses associated with human disease[a]

Alphavirus	Disease caused	Geographic distribution
Chikungunya(CHIK)	Arthropathy	Africa, Asia
Mayaro (MAY)	Arthropathy	South America
O'nyong-nyong (ONN)	Arthropathy	Africa
Igbo Ora	Arthropathy	Africa
Ross River (RR)	Arthropathy	Australia, Oceana
Sindbis Ockelbo	Arthropathy	Scandinavia
Sindbis Babanki	Arthropathy	West Africa
Barmah Forest (BF)	Arthropathy	Australia
Semliki Forest (SFV)	Systemic febrile illness	Africa, Euracia
Venezuelan equine encephalitis (VEE)	Systemic febrile illness	South and Central America
Everglades (EEE)	Systemic febrile illness	Florida, USA
Mucambo (MUC)	Systemic febrile illness	Brazil, Peru
Tonate (TON)	Systemic febrile illness	South America
Eastern equine encephalitis (EEE)	Encephalitis	North and South America
Western equine encephalitis (WEE)	Encephalitis	North and South America
Highlands J (HJ)	Encephalitis	USA

NOTE: [a]Alphaviruses are globally distributed and are a significant cause of encephalitis and/or arthritis in humans. Emerging viruses, such as Chikungunya and Ross River virus, are capable of causing massive epidemics in endemic areas and have the potential to emerge in other regions of the world.

cycle of infection, it is rare for the host vertebrate population to become totally immune. If it did so, no viremia would occur, and the virus's survival chances would be reduced. Thus, there exists a balance between vector breeding, life span, and the birth of more susceptible hosts. Continuous transmission of alphaviruses only occurs in tropical/subtropical regions and is tenuous for survival.

In the basic alphavirus maintenance cycle, neither vertebrate host nor vector suffers serious ill effects. However, explosive epidemics do occur in humans and livestock, typically coinciding with heavy rain, and concomitant mosquito population explosions. The infection cycle may be summarized as follows:[2] (1) An arthropod takes a blood meal from an infected host. There is a threshold barrier to infection, that is, the blood meal must contain above a certain concentration of virus to enable its escape from the gut, and enable infection of other tissues. (2) An incubation period of days to weeks follows, during which time virus replication occurs in the arthropod until a threshold virus titer is reached in the salivary glands. After this time, the salivary glands become persistently (noncytolytically) infected, and the virus can be transmitted to susceptible hosts for the remainder of the arthropod's life. (3) Infection of vertebrate host leads to virus multiplication, and possibly viremia (existence of virus particles in the blood), with clinically manifest symptoms. The likelihood of an epidemic depends on the infection threshold of vectors, the availability of susceptible hosts, and the level of viremia generated therein.

The incubation period in the vertebrate host prior to onset of viremia ranges from hours to days. Viremia is generally of short duration, and is at the peak level required to transmit virus to vectors only for some of this time.[3] As viremia subsides, neutralizing antibodies appear, which may persist for the lifetime of the animal.[2]

Old World Alphaviruses

In general, alphaviruses can be divided into New World and Old World alphaviruses. New World alphaviruses (e.g., Eastern Equine Encephalitis virus [EEEV] and Venezuelan Equine Encephalitis virus [VEEV]) cause more serious disease in humans than their Old World counterparts. They have the central nervous system as a clear target, causing encephalitis, and degeneration and necrosis of neurons.[4] Notable exceptions are the Fort Morgan[5] and Aura viruses,[6] which do not cause any clinical symptoms in humans. In contrast, human symptoms of Old World alphaviruses (e.g., Ross River virus [RRV], Chikungunya virus [CHIKV], and Sindbis virus [SINV]) typically include rash, headache, fever, viremia, arthritis, myositis (muscle inflammation) arthralgia (joint pain), and myalgia (muscle pain).[7] Viremia is usually short-lived, but arthralgia and myalgia can persist for several months following infection. Viral arthritides are characterized by a mononuclear infiltrate into the joint,[8] perhaps indicating that immunopathological mechanisms are more important in pathogenesis than direct virus-mediated cell injury. There is mounting evidence to suggest that alphaviruses can cause persistent infection in patients and this is further discussed in the relevant sections below.

Sindbis and Semliki Forest Viruses

The SINV and Semliki Forest virus (SFV) have been extensively studied as a model system for enveloped viruses in North America and Europe, respectively. Although rare, infection causes joint pain, headache, and weakness.[9] With one exception (a laboratory acquired infection), infection is never fatal.[10] SINV is found in Australia, Africa, India, Asia, and the Philippines. SFV is found in Africa, India, Malaysia, Vietnam, and the Philippines.[2]

O'Nyong nyong and Chikungunya Viruses

O'Nyong nyong (ONNV) and Chinkungunya (CHIKV) viruses have caused widespread epidemics of debilitating joint pain in Africa and Asia.[11] An outbreak of ONNV in 1959–1961 began in Uganda and spread over Eastern Africa, affecting more than 2 million people.[12] Furthermore, in 2006

CHIKV outbreaks in India and the Reunion Island affected more than 100,000 and 250,000 people, respectively.[13,14] *Aedes funestes, A. gambiae*, and other mosquito species are vectors for ONNV, whereas *A. aegypti* is the major vector for CHIKV. Symptoms include sudden onset of fever, headache, joint pains, rash, and lymphadenitis (inflammation of lymph nodes). ONNV has been shown to be a subtype of CHIKV.[15]

Ross River and Barmah Forest Viruses

RRV and Barmah Forest virus (BFV) are alphaviruses that cause between 5,000 and 8,000 cases of epidemic polyarthritis (EPA) annually in Australia; BFV is responsible for approximately 10% of these.[16] RRV and BFV are endemic to Australia, although there is significant risk of spread to other islands of the South Pacific region.[17,18] The closest-related viruses to RRV are SFV and SINV. The RRV polyprotein is 75% and 48% homologous to SFV and SINV polyproteins, respectively.[19]

EPIDEMIOLOGY AND HOST RANGE OF RRV

In Australia RRV appears to persist in *Aedes* mosquitos and the dominant vector varies in different regions. For example, *Ae. vigilax* is the dominant vector in tropical regions, whereas *Ae. camptorhynchus* is the dominant vector in cooler southern regions, and *Ae. annulirostris* being the main vector in inland regions.[19] Vertebrate host range is limited to placental and marsupial mammals (kangaroos and wallabies). Other native mammals, the New Holland mouse and flying foxes, for instance, are also found to be hosts for the virus.[19] Naturally acquired antibodies are found in most mammals, but most often no sign of infections can be found. Other than humans and mice, horses seem to be the only animal that shows signs of infection, which is distributed as lameness and nervous disturbances. Antibodies are rarely found in birds, which do not seem to be involved in the amplification cycles of RRV.[19]

RRV is endemic in Australia as well as Papua New Guinea.[20] RRV caused a large epidemic in 1979 in Fiji, New Caledonia, Samoa, and the Cook Islands.[20] Between 1992 and 2006, more than 55,000 cases of RRV infection were reported by the National Notifiable Diseases Surveillance System (TABLE 2). It is the leading cause of arboviral disease in Australia. Most cases of RRV infection in Australia are reported in the summer/autumn period along the coastal areas.

ROSS RIVER VIRUS DISEASE

In 1928 Nimmo recognized an "unusual epidemic" of rash and temporary arthritis, affecting 105 soldiers stationed in tropical regions of Australia.[21]

TABLE 2. Number of Ross River virus infections reported in Australian state/territory (2000–2006)[a]

Year	ACT	NSW	NT	Qld	SA	Tas	Vic	WA	Total
2000	16	750	145	1,481	416	8	319	1,089	4,224
2001	10	716	225	1,568	141	13	351	202	3,226
2002	0	180	63	886	42	117	37	131	1,456
2003	1	494	121	2,518	33	4	16	663	3,850
2004	6	699	233	2,006	53	20	91	1,102	4,210
2005	6	585	210	1,068	155	5	97	311	2,437
2006	10	1,198	267	2,597	301	13	198	784	5,368

NOTE: [a]Data courtesy of the Commonwealth Department of Health, Communicable Diseases Network, Australia New Zealand National Notifiable Diseases Surveillance System (correct as of December 2006). ACT = Australian Capital Territory; NSW = New South Wales; NT = Northern Territory; Qld = Queensland; SA = South Australia; Tas = Tasmania; Vic = Victoria; WA = Western Australia.

Hence, the disease was later termed "epidemic polyarthritis" (EPA). In 1959 Doherty and colleagues isolated an alphavirus from mosquitos (*Ae. vigilax)* near Ross River in Townsville, Queensland. They were the first to show that this virus was linked to EPA.[22]

Patients suffering from Ross River virus disease (RRVD) show the first symptoms 1–2 days after being bitten by an infected mosquito. These early symptoms include rash and pain as well as fatigue and lethargy. Body temperature remains normal in 50% of patients but in some patients body temperature can rise. Myalgia and headache are common symptoms of RRVD. Other symptoms that occur less frequently include photophobia, neck stiffness, and sore throat. Three in four patients develop joint manifestations with pain occurring while sleeping, walking, or grasping objects. Joints affected include wrists, ankles, knee joints, metacarpo, and interphalangeal joints. In about one-third of all patients arthritis occurs.[8] Rheumatic distribution is shown in wrist, knees, ankles, and finger joints, then elbows, toes, and tarsal joints. The rheumatic symptoms seem to be symmetrical in distribution. Pain is often intense, and associated with movements or in combination with swelling, heat, and redness.[8] Rash develops in two-thirds of patients and can be seen as 1–5 mm in size on trunk and limbs, and less frequently on face, scalp, palms, soles, and digital webs. About 50% of patients recovered within 6 months after onset and 95% of patients recover from disease with no symptoms after 48 months postinfection.[8] However, in some cases, symptoms have been reported to persist for up to 3 years.[8] A recent survey of RRVD patients found that RRVD could be severe at onset with level of disability that is comparable to patients with osteoarthritis, Lyme disease, or chronic rheumatoid arthritis.[23] Interestingly, RRVD in children is often mild or asymptomatic.[1] Children are well known to produce less of the arthrogenic cytokines (TNF-α and IFN-γ), a phenomenon also observed in young mice,[24] and are therefore less susceptible to a range of immunopathological and autoimmune phenomena that depend on these cytokines (rheumatoid arthritis, glandular fever, etc.).

IMMUNOLOGY OF ROSS RIVER VIRUS DISEASE

Destructive and inflammatory effects of viral cytophatic effect (CPE) and cellular immune response to viral antigen play an important role in rheumatism. Both natural killer (NK) cells and macrophages have been detected within synovial exudates from the knees of RRV-infected patients.[25] In addition, $CD4^+$ and $CD8^+$ T lymphocytes were also detected in synovial tissue sections generated from knee joint biopsies.[25] No B cells or neutrophils were found.[25]

T cells play important roles in virus infections. $CD8^+$ T cells in particular are responsible for killing virus-infected cells by inducing cytolytic mechanisms as well as producing a variety of antiviral factors. $CD4^+$ T cells were found in great numbers in synovial effusions of infected patients as well as $CD8^+$ T cells in the skin of patients suffering from EPA.[25,26] T cells may be responsible for cytotoxic elimination of the virus, as no correlation between RRV infection and antibody level could be detected.[26] Linn and colleagues (1998)[27] showed an increase of $CD8^+$ T cells in mice, which had been injected with RRV capsid protein, followed by a complete clearance of the virus[26] suggesting a requirement for $CD8^+$ T cells for mediating viral clearance. The role of $CD4^+$ T cells is less clear but may play a role in the production of host factors required for viral clearance.

Histological studies showed that monocytes and macrophages are found in high numbers compared with lymphocytes in early effusions. Viral antigens have been detected in association with the synovial macrophages of some patients with EPA.[25,28] Macrophages, but not monocytes, could be infected easily with RRV and they are the main cell type to be infected by RRV.[29] Using our novel mouse model of viral arthritis, we have shown that macrophages contribute toward the damage to the striated muscle in the hind-limbs of mice post-RRV infection.[30] A report by Soden *et al.* has used molecular techniques to detect RRV ribonucleic acid (RNA) in synovial tissue from a cohort of EPA patients more than 1 month after the onset of symptoms, providing evidence of persistent infection in the inflamed synovium.[31] It is feasible that the long-term presence of RRV and/or RRV antigens is required in the synovium for the chronic arthritis/arthralgia symptoms seen in some EPA patients. In this regard, it is interesting to note that murine macrophage-like cell cultures continued to harbor the virus for more than 50 days postinfection with no signs of CPE.[32,33] The persistent RRV infection triggered a dysregulation of various immune molecules, such as CD80 involved in antigen presentation, and IFN-γ and TNF-α.[33] Studies in our laboratory provide some evidence *in vitro* on the underlying mechanisms responsible for this persistence. We have shown that after antibody-dependent enhancement (ADE) of infection in macrophages, RRV suppressed antiviral factors,[34,35] and that macrophages infected persistently with RRV undergo variations in the expression of key soluble mediators.

Cytokines and chemokines are soluble proteins produced by immune and nonimmune cells to regulate the host immune responses.[36,37] The release of these soluble mediators following infection leads to cell activation or suppression, cell growth and survival, MHC expression, production of various host proteins, and inflammation. A role for host factors in mediating disease is suggested by the detection of IFN-γ in synovial effusions (R. Fraser and A. Cunningham, personal communication) and the isolation of IFN-γ-secreting RRV-specific T cells from humans (A. Suhrbier, personal communication).

Mateo and colleagues investigated cytokine and chemokine expression of acutely RRV-infected human synovial fibroblasts as well as acutely and chronically infected murine macrophages.[29] They found that monocyte chemotactic protein 1 (MCP-1; CCL2), IL-8 (CXCL8) and macrophage inflammatory protein 2 (MIP-2; CXCL1) were highly expressed in these cells.[29] Virus-infected synovial fibroblasts showed an increase in CCL2 and CXCL8, whereas infected murine macrophages showed increased levels of CXCL1 and CXCL8 expression after acute infection. Persistently infected macrophages that had previously cleared RRV were infected again, and these cells showed an upregulation of CXCL1, CXCL8, CCL2, and interferon inducible protein 10 (IP-10; CXCL10) mRNA compared with the noninfected controls.[29] The contrast in gene-activation profiles between acute and chronically infected macrophages supports the observation of EPA being episodic in some patients, with the potential maintenance of RRV in joint cells through periods of no disease.

Considering the fact that IFN-γ has been shown to upregulate the expression of the chemokine CCL2, it is possible that both these soluble mediators play an important role in RRV-induced inflammatory disease.

Antibody-Dependent Enhancement of Infection and Viral Arthritides

ADE of infection has been recognized as an infection strategy for arboviruses. RRV can infect macrophages via a normal cell receptor (e.g., integrin), as well as utilize ADE mechanisms to significantly enhance its infectivity on cells bearing Fc receptor (e.g., monocyte and macrophages). The reality of ADE in animals and humans cannot be ignored and is suspected as a factor for the development of severe disease. Studies with caprine arthritis–encephalitis virus (CAEV) infection in goats suggest a potential link between ADE infection and disease severity.[37] CAEV induces a gradual development of debilitating arthritis in goats, and goats vaccinated against CAEV developed severe arthritis.[38] A subsequent study has shown that goat macrophages, which infiltrate joint tissues after infection, express high levels of CXCL8 and CCL2 mRNA transcripts.[39] Previous work in our laboratory has provided insight into the signaling events associated with ADE using an RRV model of infection on murine macrophages. We found specific suppression of the antiviral factors in RRV–ADE-infected macrophage cultures that corresponded with elevated levels of IL-10 expression.[34] Furthermore, CCL2 protein was found to be produced at significantly higher levels of post-RRV–ADE infection when compared with

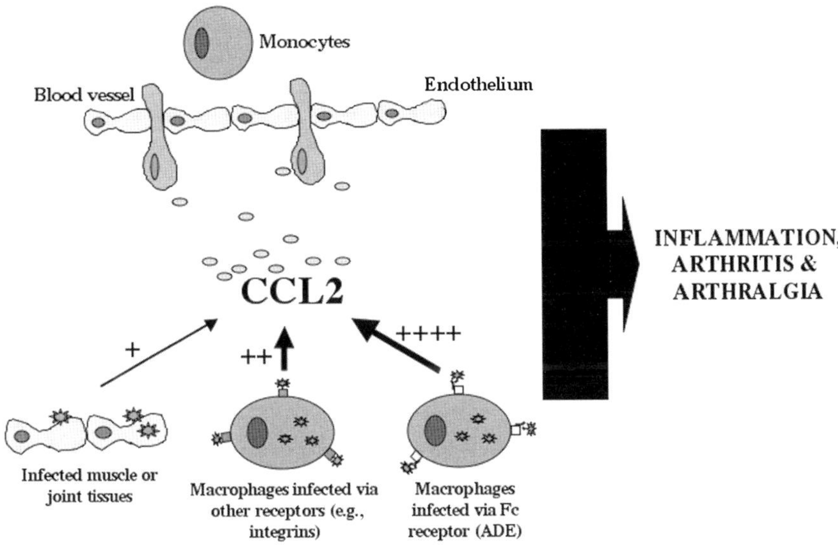

FIGURE 1. Proposed model for the role of CCL2 in RRVD. RRV infection of macrophages and cells lining tissues (e.g., muscle, joint) triggers the production of CCL2, which mediates and amplifies the recruitment of monocytes from the blood to tissue sites. These events contribute to inflammation, arthritis, and arthralgia. RRV infection via the ADE pathway results in elevated levels of CCL2 production (Rulli and Mahalingam, unpublished data).

that of non-ADE infection (N. Rulli and S. Mahalingam, unpublished data) (FIG. 1). These observations provide firm evidence to link virus–ADE infection to chemokine/cytokine dysregulation as a basis for infectious arthritic disease.

RRV-INDUCED ARTHRITIS

Understanding the mechanisms leading to virus-mediated arthritis in humans has been hampered by the absence of an appropriate animal model of disease. Direct viral replication within joint tissues is thought to lead to arthritis/arthralgia. This possibility is supported by studies of patients suffering from rubella- or RRV-induced arthritis. Rubella virus has been isolated from the joints of individuals with arthritic disease,[40] while RRV has been detected in joint effusions by antigen staining and reverse transcriptase polymerase chain reaction (RT-PCR).[41] It is possible that alphavirus-induced arthritis/arthralgia is due to the ability of the virus to replicate and mediate tissue damage within the joint. It is also likely that the host immune response plays a significant and important role in the pathogenesis of RRV inflammatory disease. As mentioned in the preceeding sections, the RRV-induced synovial effusions in affected humans are predominantly mononuclear in nature with the major cell types being monocyte/macrophages and $CD4^+$ T cells. A role for specific cytokines and chemokines in mediating disease is suggested by the detection of IFN-γ in synovial effusions (Fraser and Cunningham, personal communication) and the finding that RRV-infected synovial fibroblasts produce CCL2. Considering that RRV can infect and persists in macrophages, it is feasible that the long-term presence of RRV and/or RRV antigens is required in the synovium for the chronic arthritis/arthralgia symptoms seen in some EPA patients. However, using an animal model, a direct role for various immune cells (e.g., $CD4^+$ T cells) or soluble proteins (e.g., IFN-γ or CCL2) in RRV inflammatory disease will be forthcoming.

Mouse Model of RRV Arthritis/Arthralgia

Early mouse studies provided evidence for muscle inflammation and associated damage as being the likely physiological basis of RRV-induced symptoms in mice.[41] However, these early studies did not find evidence for viral replication or virus-mediated joint pathology in joint tissues. In recent years, our laboratories (in collaboration with Dr Mark Heise, University of North Carolina) have established a mouse model of RRV-induced inflammatory disease, which represents one of the first models of viral arthritis. The model has many characteristics in common with alphavirus-induced arthritis/athralgia in humans and is an excellent system for studying viral arthritides.

Following RRV infection, mice developed disease signs such as weight loss and hind limb weakness by day 5.[42] Peak disease occurred between days 7 and 10 after infection. The most severe disease manifestation was the complete loss of limb function. However, these mice were able to regain complete use of their limbs within 15 days of onset of the disease. Virus was detectable in joint-associated tissues, including ligaments and periosteum. Cells infected with RRV were frequently observed in the synovial cavity of the joint. Viral replication was also prominent within skeletal muscle. Inflammation was detected by day 5 after infection and peaked between days 7–10 p.i. in the muscle (FIG. 2) and joints (FIG. 3). Improved limb function in the infected animals corresponded with the resolution of the inflammation.

The types of cellular infiltrates isolated from the joint and muscle tissues of mice were found to be consistent with the observation found in synovial exudates of patients with RRVD. Significant increases in percentage and total number of NK cells and inflammatory macrophages within the skeletal muscle tissue were also detected.[42] Increased numbers of both $CD4^+$ and $CD8^+$ T cells were also detected. Clearly, this mouse model of RRV-induced inflammatory disease has allowed us to begun to unravel the pathobiology of viral-induced joint and muscle inflammation. For instance, using our novel mouse model of RRVD, we have identified the critical role for macrophages in mediating viral-induced tissue damage and inflammation. The selective depletion of

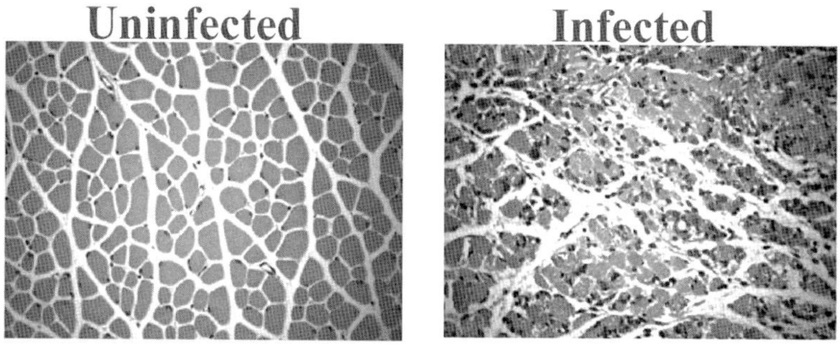

FIGURE 2. Ross River virus induces inflammation in hind limb skeletal muscle tissue of mice. C57BL/6J mice were infected with 10^3 plaque forming unit (pfu) of RRV by subcutaneous injection in the left-rear footpad. At 7 days postinfection mice were perfused with 4% paraformaldehyde and 5 μM thick paraffin-embedded sections generated from the quadriceps muscle were H & E stained. Magnification=200×.

macrophages prior to RRV infection in our mouse model abolished the *in situ* tissue damage and inflammation.[30]

CONCLUDING REMARKS

It is well known in adults that RRV infection and infection with other potentially arthrogenic organisms that infection by itself is not sufficient for arthrogenesis. Other conditions need to be met. Enhanced secretion of arthrogenic cytokines is clearly one of these conditions, and we believe that enhanced

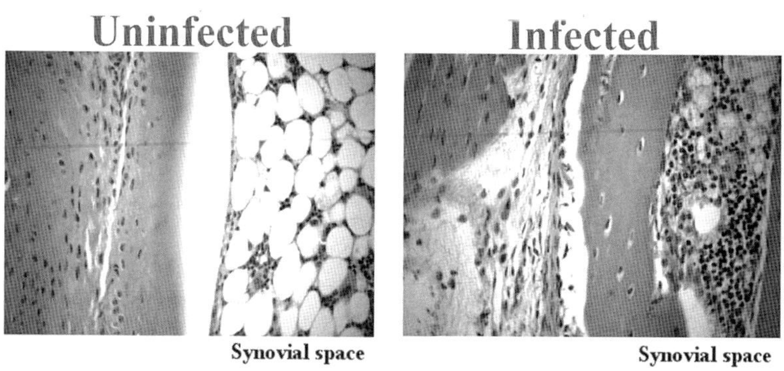

FIGURE 3. Ross River virus induces inflammation in joint tissue of mice. C57BL/6J mice were infected with 10^3 pfu of RRV by subcutaneous injection in the left-rear footpad. At 7 days postinfection mice were perfused with 4% paraformaldehyde. Following decalcification, 5-μM-thick paraffin-embedded sections from uninfected and RRV-infected animals were H & E stained. Magnification = 200×.

RRV infection mediated by ADE is likely to contribute significantly to elevated levels of these cytokines after early infection. At present treatment with nonsteroidal anti-inflammatory drugs (NSAIDs) remains the best treatment for RRVD. The low prevalence of the disease and the high cost involved in vaccine development complicates the commercial development of such vaccines.[43] Current research conducted using the mouse RRV study system will significantly enhance our understanding of the molecular and cellular pathogenesis of Togavirus-induced rheumatic disease. It is reasonable to speculate that this may ultimately lead to discovery of novel and improved therapies against these diseases.

ACKNOWLEDGMENTS

Suresh Mahalingam is a recipient of the Australian NHMRC R. Douglas Wright Fellowship. Suresh Mahalingam acknowledges the Australian National Health and Medical Research Council for their support (project grant 303404).

REFERENCES

1. Johnston, R.E. & C.J. Peters. 1996. Alphaviruses. *In* Fields Virology, 3rd ed., Vol 1. B.N. Fields, D.M. Knipe, P.M. Howley, *et al.* Eds.: 843–898. Lippincott-Raven. Philadelphia.
2. Chamberlain, R.W. 1980. Epidemiology of arthropod-borne togaviruses. *In* The Togaviruses: Biology, Structure, Replication. R.W. Schlesinger, Ed.: 175–227. Academic Press Inc. New York, London, Sydney.
3. Schlesinger, R.W. 1980. Virus-host interactions in natural and experimental infections with alphaviruses and flaviviruses. *In* The Togaviruses: Biology, Structure, Replication. R.W. Schlesinger, Ed.: 83–106. Academic Press Inc. New York, London, Sydney.
4. Shope, R.E. 1980. Medical siginficance of togaviruses: an overview of diseases caused by togaviruses in man and in domestic and wild vertebrate animals. *In* The Togaviruses: Biology, Structure, Replication. R.W. Schlesinger, Ed.: 47–82. Academic Press Inc. New York, London, Sydney.
5. Calisher, C.H., T.P. Monath, D.J. Muth, *et al.* 1980. Characterization of Fort Morgan virus, an alphavirus of the Western equine encephalitis virus complex in an unusual ecosystem. Am. J. Trop. Med. Hyg. **29:** 1428–1440.
6. Rumenapf, T., E.G. Strauss, J.H. Strauss. 1995. Aura virus is a new world representative of Sindbis-like viruses. Virology **208:** 621–633.
7. Rulli, N.E., A. Suhrbier, L. Hueston, *et al.* 2005. Ross River virus: molecular and cellular aspects of disease pathogenesis. Pharmacol. Therapeut. **107:** 329–342.
8. Fraser, J.R. 1986. Epidemic polyarthritis and Ross River virus disease. Clin. Rheum. Dis. **12:** 369–388.
9. Mathiot, C.C., G. Grimaud, P. Garry, *et al.* 1990. An outbreak of human Semliki Forest virus infections in Central African Republic. Am. J. Trop. Med. Hyg. **42:** 386–393.

10. WILLEMS, W.R., G. KALUZA, C.B. BOSCHEK, *et al.* 1979. Semliki Forest virus: cause of a fatal case of human encephalitis. Science **203:** 1127–1129.
11. TUPANCESKA, D., A. ZAID, N.E. RULLI, *et al.* 2007. Ross River virus: an arthritogenic alphavirus of significant importance in the asia pacific. *In* Emerging Viral Disease of Southeast Asia. Vol 4. S.K. Lal, Ed.: 94–111. Issues Infect Dis. Karger Basel, Switzerland.
12. HADDOW, A.J., C.W. DAVIES & A.J. WALKER. 1960. O'nyong nyong fever: an epidemic virus disease in West Africa. Trans. R. Soc. Trop. Med. Hyg. **54:** 517–522.
13. PASTORINO, B.F., J.J. MUYEMBE-TAMFUM, M.F. BESSAUD, *et al.* 2004. Epidemic resurgence of Chikungunya virus in democratic republic of the Congo: identification of a new central African strain. J. Med. Virols. **74:** 277–282.
14. SCHUFFENECKER, I., I. ITEMAN, A. MICHAULT, *et al.* 2006. Genome microevolution of chikungunya viruses causing the Indian Ocean outbreak. PLOS Med. **3:** 1058–1070.
15. CALISHER, C.H., R.E. SHOPE, W. BRANDT, *et al.* 1980. 1980b. Proposed antigenic classification of registered arboviruses I. Togaviridae, Alphavirus. Intervirology **14:** 229–232.
16. MACKENZIE, J.S. & D.W. SMITH. 1996. Mosquito-borne viruses and epidemic polyarthritis. Med. J. Aust. **164:** 90–93.
17. KELLY-HOPE, L.A., B.H. KAY, D.M. PURDIE & G.M. WILLIAMS. 2002. The risk of Ross River and Barmah Forest virus disease in Queensland: implications for New Zealand. Aust. N. Z. J. Public Health **26:** 69–77.
18. LIDBURY, B., S. MAHALINGAM. 2004. A role for chemokine activity in alphavirus pathogenesis: evidence from the analysis of polyarthritis and myalgia post Ross River virus infection. *In* Chemokines in Viral Infections. S. Mahalingam, Ed.: 92–106. Medical Intelligence Unit, Landes Biosciences. Austin, TX.
19. RUSSELL, R.C. 2002. Ross River virus: ecology and distribution. Ann Rev Entomol **47:** 1 31.
20. AASKOV, J.G., J.U. MATAIKA, G.W. LAWRENCE, *et al.* 1981. An epidemic of Ross River virus infection in Fiji, 1979. Am. J. Trop. Med. Hyg. **30:** 1053–1059.
21. NIMMO, J.R. 1928. An unusual epidemic. Med. J. Aust. **1:** 549–550.
22. DOHERTY, R.L. 1977. Arthropod-borne viruses in Australia, 1973–1976. Aust. J. Exp. Biol. Med. Sci. **55:** 103–130.
23. MYLONAS, A.D., A.M. BROWN, T.L. CARTHEW, *et al.* 2002. Natural history of Ross River virus-induced epidemic polyarthritis. Med. J. Aust. **177:** 356–360.
24. ORDEMANN, R., R. HUTCHINSON, J. FRIEDMAN, *et al.* 2002. Enhanced allostimulatory activity of host antigen-presenting cells in old mice intensifies acute graft-versus-host disease. J. Clin. Invest. **109:** 1249–1256.
25. FRASER, J.R., V.M. RATNAMOHAN, J.P. DOWLING, *et al.* 1983. The exanthem of Ross River virus infection: histology, location of virus antigen and nature of inflammatory infiltrate. J. Clin. Pathol. **36:** 1256–1263.
26. FRASER, J.R. 1984. Mononuclear cell types in chronic synovial effusions of Ross River virus disease. Aust. N. Z. J. Med. **14:** 505–506.
27. LINN, M.L., L. MATEO, J. GARDNER & A. SUHRBIER. 1998. Alphavirus-specific cytotoxic T lymphocytes recognize a cross reactive epitope from capsid protein and can eliminate virus from persistently infected macrophages. J. Virol. **72:** 5146–5153.

28. HAZELTON, R.A., C. HUGHES & J.G. AASKOV. 1985. The inflammatory response in the synovium of a patient with Ross River arbovirus infection. Aust. N. Z. J. Med. **15:** 336–339.
29. MATEO, F., M.L. LINN, S.R. MCCOLL, *et al.* 2000. An arthrogenic alphavirus induces Monocyte Chemoattractant Protein-1 and Interleukin-8. Intervirology **43:** 55–60.
30. LIDBURY, B.A., C. SIMEONOVIC, G.E. MAXWELL, *et al.* 2000. Macrophage-induced muscle pathology results in morbidity and mortality for Ross River virus-infected mice. J. Infect. Dis. **181:** 27–34.
31. SODEN, M., H. VASUDEVAN, B. ROBERTS, *et al.* 2000. Detection of viral ribonucleic acid and histologic analysis of inflamed synovium in Ross River virus infection. Arthritis Rheum. **43:** 365–369.
32. LINN, M.L., J.G. AASKOV & A. SUHRBIER. 1996. Antibody-dependent enhancement and persistence in macrophages of an arbovirus associated with arthritis. J. Gen. Virol. **77:** 407–411.
33. WAY, S.R., B.A. LIDBURY & J.L. BANYER. 2002. Persistent Ross River virus infection of murine macrophages: an *in vitro* model for the study of viral relapse and immune modulation during long-term infection. Virology **301:** 281–292.
34. MAHALINGAM, S., B.A. LIDBURY. 2002. Suppression of LPS-induced antiviral transcription factor (STAT1 and NF-κB) complexes by antibody-mediated enhancement of Ross River virus in macrophages. Proc. Natl. Acad. Sci. U.S.A. **99:** 13819–13824.
35. LIDBURY, B.A., S. MAHALINGAM. 2000. Specific ablation of antiviral genes in macrophages by antibody dependent enhancement of Ross River virus infection. J. Virol. **74:** 8376–8381.
36. MAHALINGAM, S., J. FRIEDLAND, M. HEISE, *et al.* 2003. Chemokines and viruses: friends or foes. Trends Microbiol. **11:** 383–391.
37. RAMSHAW, I.A., A.J. RAMSAY, G. KARUPIAH, *et al.* 1997. Cytokines and immunity to viral infections. Immunol. Rev. **159:** 119–135.
38. MCGUIRE, T.C., D.S. ADAMS, G.C. JOHNSON, *et al.* 1986. Acute arthritis in caprine arthritis-encephalitis virus challenge exposure of vaccinated or persistently infected goats. Am. J. Vet. Res. **47:** 537–540.
39. LECHNER, F., E. PETERHANS, D.A. DOBBELAERE, *et al.* 1997. Caprine arthritis encephalitis virus dysregulates the expression of cytokines in macrophages. J. Virol. **71:** 7488–7497.
40. OGRA, P.L. & J.K. HERD. 1971. Arthritis associated with induced rubella infection. J. Immunol. **107:** 810–813.
41. MURPHY, F.A., W.P. TAYLOR, C.A. MIMMS & I.D. MARSHALL. 1973. Pathogenesis of Ross River virus infection in mice. I. muscle, heart, and brown fat lesions. J. Infect. Dis. **127:** 129–138.
42. MORRISON, T., A. WHITMORE, R. SHABMAN, *et al.* 2006. Characterization of Ross River virus tropism and virus-induced inflammation in a mouse model of viral arthritis and myositis. J. Virol. **80:** 737–749.
43. SUHRBIER, A. & M.L. LINN. 2004. Clinical and pathological aspects of arthritis due to Ross River virus and other alphaviruses. Curr. Opin. Rheumatol. **16:** 374–379.

Rapid Identification of Emerging Infectious Agents Using PCR and Electrospray Ionization Mass Spectrometry

RANGARAJAN SAMPATH, THOMAS A. HALL, CHRISTIAN MASSIRE, FENG LI, LAWRENCE B. BLYN, MARK W. ESHOO, STEVEN A. HOFSTADLER, AND DAVID J. ECKER

Ibis Biosciences, Inc., a Subsidiary of Isis Pharmaceuticals, Carlsbad, California, USA

ABSTRACT: Newly emergent infectious diseases are a global public health problem. The population dense regions of Southeast Asia are the epicenter of many emerging diseases, as evidenced by the outbreak of Nipah, SARS, avian influenza (H5N1), Dengue, and enterovirus 71 in this region in the past decade. Rapid identification, epidemiologic surveillance, and mitigation of transmission are major challenges in ensuring public health safety. Here we describe a powerful new approach for infectious disease surveillance that is based on polymerase chain reaction (PCR) to amplify nucleic acid targets from large groupings of organisms, electrospray ionization mass spectrometry (ESI-MS) for accurate mass measurements of the PCR products, and base composition signature analysis to identify organisms in a sample. This approach is capable of automated analysis of more than 1,500 PCR reactions a day. It is applicable to the surveillance of bacterial, viral, fungal, or protozoal pathogens and will facilitate rapid characterization of known and emerging pathogens.

KEYWORDS: PCR; RT-PCR; ESI-MS; mass spectrometry; infectious disease surveillance; viral diseases; broad detection; emerging infections

INTRODUCTION

Southeast Asia has been the source of several emerging zoonotic- and vector-borne disease outbreaks over the years.[1] These include acute encephalitis caused by a novel paramyxovirus, Nipah virus;[2,3] hand, foot, and mouth disease (HFMD) caused by coxsackie A16 and enterovirus 71;[4] viral encephalitis and

Address for correspondence: Rangarajan Sampath, Ibis Biosciences, Inc., a subsidiary of Isis Pharmaceuticals, 1891 Rutherford Rd., Carlsbad, CA 92008, USA. Voice: 760-603-2652; fax: 760-603-4653.
rsampath@ibisbio.com

Ann. N.Y. Acad. Sci. 1102: 109–120 (2007).
doi: 10.1196/annals.1408.008

hemorrhagic fevers caused by arboviruses, such as Dengue and Japanese encephalitis viruses;[5] and episodic outbreaks of influenza A virus (H5N1, H9N2, etc.) and acute respiratory syndrome caused by the SARS coronavirus.[6–9]

A fundamental problem in rapid identification of the etiology of infectious disease outbreaks is the sheer number of microbes, both known and novel, that could potentially be the causative agents. According to a recent review,[10] there are more than 1,400 species of microbes known to be pathogenic to humans. Of these, more than 60% were found to be zoonotic and more than 175 pathogenic species were associated with emerging infectious diseases. These numbers do not include numerous strain variants of each organism or classes of pathogens that infect plants or animals. Further, species variants of an otherwise benign group of microbes may prove infectious to humans and strain-level genetic changes may confer resistance to antimicrobial agents.

Most of the new technologies being developed for detection of infectious agents incorporate a version of quantitative polymerase chain reaction (PCR), based upon the use of highly specific primers and probes and designed to selectively identify specific pathogenic organisms. Paradoxically, this approach requires assumptions about the type and strain of the infectious agent, limiting detection to a small number of known pathogens. It is both impractical and cost prohibitive to run a large number of individual tests in an attempt to identify every known infectious organism. Moreover, existing tests for known organisms will not identify newly emerging organisms.

Broad-range PCR methods provide an alternative to single-agent tests. By amplifying gene targets conserved across groups of organisms, broad-range PCR has the potential to generate amplification products across entire genera, families, or, as with bacteria, an entire domain of life. This strategy has been successfully employed in the past for bacterial and viral detection.[11–16] The drawback of this approach for epidemiologic applications is that the analysis of PCR products for mixed amplified samples requires sequencing of hundreds of colonies per reaction, and thus it cannot be performed rapidly or cost effectively on large numbers of samples. New approaches to the parallel detection of multiple infectious agents include multiplexed PCR methods[17,18] and microarray strategies.[19–21] Microarray strategies are promising because undiscovered organisms might be detected by hybridization to probes designed to conserved regions of known families of bacteria and viruses. However, these methods are not well suited for the high-throughput detection of a broad spectrum of microorganisms potentially present in complex biological mixtures.

Triangulation identification for genetic evaluation of risks (TIGER) is a universal pathogen detection platform developed under a Defense Advanced Research Projects Agency (DARPA) biodefense program that is capable of identifying an extraordinarily broad range of pathogens (FIG. 1). The process uses electrospray ionization mass spectrometry (ESI-MS) and base composition analysis of PCR amplification products from highly conserved genomic regions to identify and determine the relative quantity of pathogenic bacteria,

FIGURE 1. Automated PCR/ESI-MS biosensor. Shown in this view are key modules including amplicon purification (desalting), plate stackers, sample injection, and the mass spectrometer. Precise molecular weight determinations of amplicons yield unambiguous base compositions that are used to uniquely "fingerprint" each pathogen. The automated system is capable of analyzing more than 1,500 PCR reactions in 24 h.

viruses, fungi, or protozoa present in a sample. The fundamental distinction between this PCR/ESI-MS strategy and traditional (specific) nucleic acid tests is the nature of the question being answered. Traditional nucleic acid tests answer the question: "Is infectious organism X in my sample?" The PCR/ESI-MS method answers the question: "What infectious organisms are in my sample?" In effect, this is equivalent to running many thousands of specific nucleic acid tests because the identity of the infectious organism being detected does not need to be anticipated. The premise of this technology is that it can provide rapid, sensitive, and cost-effective detection of a broad range of "normal" pathogenic organisms and simultaneously identify unexpected or emerging infectious organisms. This broad function technology may be the only practical way to rapidly diagnose diseases caused by emerging infectious organisms that might otherwise be missed or mistaken for a more common infection.[22] The technology has been validated for a variety of applications, including emerging virus identification,[23] determination of the pathogens involved in acute pneumonia epidemics,[24] and identification of biological weapons agents in a variety of sample types.[25,26]

The basis of the above approach is the principle that, despite the enormous diversity of microbes, there are sets of essential common features in their biomolecules. Parallel detection of infectious agents takes advantage of highly conserved regions within genes that flank highly variable regions. These conserved genomic features serve as anchors for broad-range PCR priming to amplify, without prejudice, nucleic acids from most microorganisms in a sample. Such broad-range PCR reactions targeted across the broadest possible grouping of organisms typically result in a mixture of amplicons, reflecting the complexity of the starting sample. With no further separation, this mixture of PCR amplicons is desalted and analyzed by ESI-MS for accurate mass measurement, enabling us to determine the base composition (numbers of A, G, C, and T nucleotides) of the PCR amplicons.[23,25,26] Amplification relies on highly conserved regions, but the variability in sequence between the two primer sites, reflected in the base composition, allows us to distinguish between organisms.

A key aspect of this approach is the concept of "triangulation." In this context, triangulation refers to taking measurements from multiple loci distributed across the microbial genome and using the combination of the measured base composition signatures to identify the organism. While no single PCR target region might give definitive species resolution for all microbes, species can be clearly identified and differentiated from one another using the triangulation strategy across multiple loci. A database of expected base compositions for each primer region for all known organisms was generated using an *in silico* PCR search algorithm (electronic PCR or ePCR). An existing ribonucleic acid (RNA) structure search algorithm has been modified to include PCR parameters, such as hybridization conditions, mismatches, and thermodynamic calculations.[27] This also provides information on primer specificity of the selected primer pairs. The regions are selected so that the nucleic acid composition of the resulting amplicons has sufficient variability to maximally resolve the target organisms. Thus, the measured base compositions from strategically selected locations of the genome are used as a signature to identify and distinguish the organisms present in the original sample.

Detailed experimental protocols have been published elsewhere.[25] Microbes have been identified from human and animal samples (throat swabs, skin wipes, hair, tissue) and from samples obtained from the environment (food products, soil, water, surface wipes, air). Nucleic acids are extracted from bacteria using either a heat or mechanical lysis step. Viral genome isolation is performed using commercially available kits that have been optimized for use in the system. Samples are divided into wells of a microtiter plate that contain pairs of broad-range PCR primers.

A key element of the PCR/ESI-MS system is the ability to quantify the levels of genomic material present in the samples. To accomplish this, we include an internal calibrant in each PCR reaction performed. The calibrant consists of a specially designed nucleic acid sequence that is similar to, but distinguishable from, sequences of organisms we might detect. This allows for the simultaneous presence of an accurately quantified amount of a known calibrant deoxyribonucleic acid (DNA) with similar amplification properties to the target

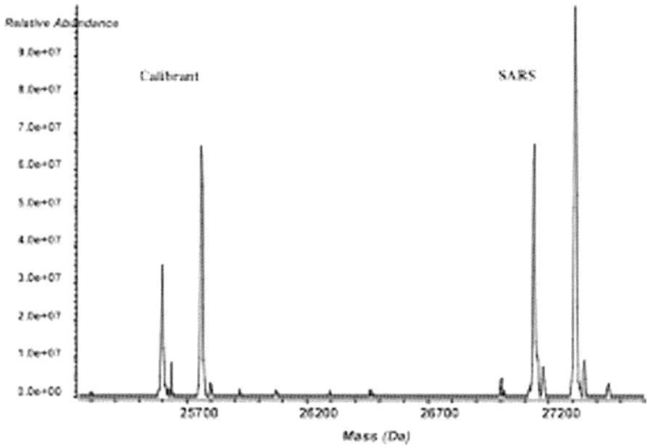

FIGURE 2. Mass spectra of calibrant and SARS coronavirus (SARS CoV) samples are shown. Both strands were amplified and thus there are two major peaks for both calibrant and SARS. The two templates were observed to be at a relative abundance ratio of approximately 1:1.5. From the input number of calibrant nucleic acid molecules (3×10^4), we estimated that SARS RNA was present in the sample at $\sim 4.5 \times 10^4$ genome copies.

sequence during the PCR cycle. Further, this serves as an internal positive control for every PCR reaction. The relative abundances of the end products (as measured by the mass peak heights) can be used in conjunction with the starting concentration of the calibrant to accurately estimate the amount of target sequence in the starting sample (FIG. 2). These measurements are highly reproducible over a wide range of concentrations and provide improved confidence for the determination of the quantity of the infectious agents in the sample.[24,25]

DETECTION AND DISCRIMINATION OF KNOWN AND EMERGING VIRUSES

In this section, we describe two specific applications of the PCR/ESI-MS method for identifying known and emerging viruses of particular relevance to Southeast Asia. These concepts are equally applicable to bacteria,[24,25] other viruses, fungi, and protozoa.

SARS Coronavirus

At the beginning of the SARS outbreak in early 2003, the etiology of the causative agent was not known. Several initial reports based on preliminary characterization of the isolated virus suggested it could be a member of one of several viral families, including *Coronaviridae*. We designed several broad-range primers targeting the various genera within these families. For broad-range detection of all coronaviruses, two PCR primer target regions in orf-1b, one in the RNA-dependent RNA polymerase (RdRp) and the other in Nsp14, were identified prior to publication of the SARS sequence.[23] Locations of primers within these regions were optimized both for sensitivity and broad-range priming potential simultaneously by amplifying dilutions of multiple, diverse coronaviruses. *In silico* analysis of the final primer pairs against GenBank nucleotide database sequences showed that these primers would be expected to amplify all coronaviruses but no other viruses, bacteria, or human DNA. [27]

Multiple isolates of a number of different coronavirus species were obtained from various sources (including the American Type Culture Collection [ATCC] and the Naval Health Research Center in San Diego). SARS isolates were obtained from the Centers for Disease Control and Prevention (CDC) and from Michael Buchmeier (The Scripps Research Institute, La Jolla, CA). PCR products for all isolates were generated, desalted, and analyzed by ESI-MS by methods described previously.[28] The spectral signals were algorithmically processed to obtain base compositions of sense and antisense strands of each amplicon.[29] The results from the analysis of 14 coronavirus isolates are shown in TABLE 1. For both target regions, the measured signals agreed with compositions expected from the known coronavirus sequences in GenBank. Several of the isolates used in this study did not have a genome sequence record in GenBank. Nevertheless, we were able to amplify all test viruses and experimentally determine their base compositions. These experimentally determined base compositions were later confirmed by sequencing and deposited in GenBank (AY874541, AY878317-24). Thus the strategy described here enables identification of organisms without the need for prior knowledge of their sequence.

We further demonstrated the potential to detect multiple viruses in the same sample, as might occur during a co-infection. Pooled viral extracts from three human coronavirus isolates, HCoV-229E, HCoV-OC43, and SARS CoV, were amplified in a mixture and analyzed by PCR/ESI-MS. Signals from all three viruses were clearly detected and resolved in the mass spectrum demonstrating that co-infections of more than one coronavirus species could be identified. In addition to broad identification of all coronaviruses tested, including the SARS CoV, we were able to accurately classify the SARS isolates as unrelated but distant members of the group 2 coronaviruses, consistent with the findings of Snijder *et al.*[30] These results are described in detail in Sampath *et al.*[23]

Vector-Borne Viruses

Viral encephalitidis is caused by a number of different viruses from different viral families. Important among these are members of the genus *Alphavirus*

TABLE 1. Coronaviruses and their experimentally determined base compositions. Sense and antisense strand masses are shown for each isolate

Group	Abbreviation	Source	Measured masses (Da)	Derived Base compositions	Measured masses (Da)	Derived base compositions
	CCoV 1-71	VR809	27486.514	A24 G24 C8 T32	42475.955	A33 G31 C19 T54
			26936.574	A32 G8 C24 T24	42185.117	A54 G19 C31 T33
	CCoV -TN449	VR2068	27471.510	A24 G24 C9 T31	42474.899	A34 G30 C18 T55
			26952.548	A31 G9 C24 T24	42184.072	A55 G18 C30 T34
	FCoV WSU	VR-989	27471.517	A24 G24 C9 T31	42490.945	A33 G31 C18 T55
	79-1683		26952.556	A31 G9 C24 T24	42169.118	A55 G18 C31 T33
1	FCoV DF2	VR2004	27472.497	A23 G25 C10 T30	42450.904	A33 G30 C19 T55
			26953.536	A30 G10 C25 T23	42209.081	A55 G19 C30 T33
	HCoV 229E	VR740	27450.532	A25 G24 C11 T28	42462.994	A36 G30 C20 T51
			26975.545	A28 G11 C24 T25	42198.061	A51 G20 C30 T36
	HCoV 229E	NHRC	27450.506	A25 G24 C11 T28	42462.930	A36 G30 C20 T51
			26975.512	A28 G11 C24 T25	42198.040	A51 G20 C30 T36
	BCoV Calf	VR874	27358.452	A22 G22 C12 T32	42606.039	A38 G32 C15 T52
	Diarrheal virus		27066.586	A32 G12 C22 T22	42052.897	A52 G15 C32 T38
	HCoV OC43	NHRC	27328.473	A22 G22 C14 T30	42580.959	A38 G31 C15 T53
			27098.562	A30 G14 C22 T22	42076.028	A53 G15 C31 T38
	MHV MHV1	VR261	27344.491	A21 G23 C14 T30	42602.022	A37 G34 C18 T48
			27083.564	A30 G14 C23 T21	42061.016	A48 G18 C34 T37
2	MHV JHM-	VR1426	27344.497	A21 G23 C14 T30	42529.960	A34 G34 C21 T48
	thermostable		27083.571	A30 G14 C23 T21	42136.047	A48 G21 C34 T34
	MHV MHV-	VR764	27344.503	A21 G23 C14 T30	42599.989	A34 G35 C18 T50
	A59		27083.572	A30 G14 C23 T21	42064.089	A50 G18 C35 T34
	RtCoV 8190	VR1410	27344.491	A21 G23 C14 T30	42544.967	A34 G34 C20 T49
			27083.567	A30 G14 C23 T21	42120.041	A49 G20 C34 T34
3	IBV Egg-	VR22	27396.544	A24 G24 C14 T26	42530.984	A33 G32 C17 T55
	adapted		27032.524	A26 G14 C24 T24	42129.100	A55 G17 C32 T33
	SCoV TOR2	University of	27298.518	A27 G19 C14 T28	42519.906	A34 G33 C20 T50
4		Manitoba	27125.542	A28 G14 C19 T27	42144.026	A50 G20 C33 T34
	SCoV Urbani	CDC	27298.518	A27 G19 C14 T28	42519.906	A34 G33 C20 T50
			27125.542	A28 G14 C19 T27	42144.026	A50 G20 C33 T34

(*Togaviridae* family) and the genus *Flavivirus* (*Flaviviridae* family); both are single-strand, positive-strand RNA viruses. Thcsc arc major vector-borne diseases in Southeast Asia and Pacific Rim countries. While the alphaviruses Venezuelan, and eastern and western equine encephalitis viruses (VEEV, EEEV, and WEEV) have received much attention in the Americas for their ability to infect humans and their potential to be used as biowarfare agents, Ross River, Chikungunya, Barmah Forest, and other alphaviruses are common and detrimental to public health in the Asia Pacific regions.[5,31,32] Genomic changes drive the emergence and reemergence of these viruses as high-consequence pathogens by causing alterations in host range and virulence properties,[33] and the resulting diversity at the nucleotide and protein levels poses a great challenge for identification and diagnostic strategies. As an example, VEE complex viruses display 13 serologically distinct types belonging to 6 antigenic subtypes (I to VI), and the EEE virus antigenic complex has 4 antigenic subtypes (I–IV), spanning the North and South American isolates.[34]

Our approach for identifying and discriminating these viruses was to design several primer pairs that would amplify all members of the *Alphavirus* genus, and then design primers that amplify distinct regions of the Alphavirus isolates to distinguish subtypes.[25] Broad-range primer pairs targeting the 5′-end of the RNA virus genome (nonstructural protein, nsP1 gene) were designed and tested against a number of different viral isolates, including both old world and new world alphaviruses. All of the test isolates showed a perfect match to the expected base compositions where prior data were known, and the tested viral isolates were readily distinguished from one other (TABLE 2). Importantly, the base composition data for some of the isolates represented new measurements from regions that have not been sequenced, showing the potential of this strategy to identify previously unknown or newly emerging viruses. Additional confirmation primer pairs targeting the old and new world alphaviruses have been developed and together they provide complete resolution among all known alphavirus species (Eshoo *et al.*, Personal Observation).

Influenza Virus Surveillance

The biosensor system described here is uniquely suited to influenza virus detection and surveillance in avian, human, or other animal reservoirs. Broad surveillance capability for all influenza viruses is complemented by simultaneous type and subtype level resolution of various identified influenza viruses. This will be useful in monitoring the emergence of new influenza viruses. Based upon the analysis of multiple influenza sequence alignments, pan-influenza virus PCR primer sets have been developed that are capable of amplifying all three influenza virus species (A, B, and C) and subtypes (HxNy) from different animal hosts (human, avian, swine, etc.) and to distinguish their essential molecular features using base composition signatures (Sampath *et al.*, Personal Observation). To measure the breadth of coverage and resolution offered by this panel we tested 50 different influenza virus isolates, including

TABLE 2. Alphavirus test isolates and their expected/observed base compositions.[a,b]

Virus	Strain	Sequence available	Expected base composition	Observed base composition
VEE	3908 (subtype IC, 1995)	Yes	[21 23 23 19]	[21 23 23 19]
VEE	66637 (subtype ID, 1981)	Yes	[21 23 23 19]	[21 23 23 19]
VEE	68U201 (subtype IE, 1968)	Yes	[22 25 19 20]	[22 25 19 20]
VEE	243937 (subtype IC, 1992)	Yes	[21 23 23 19]	[21 23 23 19]
WEE	OR71 (71V1658)	Yes	[22 26 19 19]	[22 26 19 19]
WEE	SD83 (R-43738)	No	Unknown	[22 26 19 19]
WEE	ON41 (McMillan)	No	Unknown	[22 27 18 19]
WEE	Fleming (Fleming)	No	Unknown	[22 25 19 20]
EEE	(Parker strain)	Yes	[23 25 19 19]	[23 25 19 19]

[a]Three of the WEE isolates shown in grey represent *de novo* measurements.
[b]All isolates were obtained from Dr. Scott Weaver, UTHSC, Galveston, TX.

22 avian isolates (representing 20 different H/N types, collected from 9 different species of wild birds), 18 human influenza A isolates (8 H1N1, 10 H3N2), 4 swine isolates (including 1 novel type), 1 equine isolate, and 6 human influenza B isolates. The avian influenza samples represented a variety of host species, influenza virus types, geographic locations, and dates of isolation. Despite the diversity of this sample set, the broad-range primers generated amplicons from all isolates, while the base composition signatures from amplicons obtained with these primers distinguished the isolates (Sampath *et al.*, manuscript submitted). The assay is high throughput (~300 samples/24 h) and rapid (sample to answer in <6 h). *In silico* analysis of >1,400 influenza virus sequences from the GenBank database showed that the primer panel would detect *all known* influenza viruses and distinguish >90% of all species, with 100% resolution between avian H5N1 and any human isolate. Several hundred human respiratory specimens have been screened using this assay and have shown >97% sensitivity and specificity of detection compared to traditional characterization.

SUMMARY

TIGER strategy uses a paradigm that allows simultaneous identification of multiple pathogens, including all bacteria, viruses, fungi, and other microorganisms present in a sample without the need for prior knowledge of a specific nucleotide target sequence. Thus previously unanticipated, newly emerging, or potentially unknown organisms can be identified using this platform. The strategy described here uses integrated ESI-MS and base composition analysis of broad-range PCR products. Broad-range PCR reactions are capable of producing products from groups of organisms, rather than single species, and the information content of each PCR reaction is very high. In many cases, including the SARS-CoV detection described above, priming across broadly conserved regions provides taxonomic resolution at the species level. Where additional subspecies level classification becomes important, broad primers can be supplemented with species-specific primers that can detect even single nucleotide changes (SNPs) at highly variable loci. As a genotypic method for microbial identification, PCR/ESI-MS can provide strain- or isolate-level information.[35] These data may be important in epidemiological monitoring to determine the origin of a particular outbreak.

Due to its high throughput capabilities, the methods described will have great utility in a routine survey and detection setting. The mass spectrometer is capable of analyzing complex PCR products at a rate of approximately 1 min per sample. Because the process is performed in an automated, microtiter plate format, large numbers of samples can be examined (>1,400 PCR reactions/day/instrument), making this process practical for large-scale analysis of clinical or environmental surveillance samples in public health laboratory settings. This approach can be extended to other viral, bacterial, fungal, or protozoal pathogen groups and is a powerful new paradigm for timely identification of previously unknown organisms that cause disease in humans or animals and for monitoring the progress of epidemics.

CONFLICT OF INTEREST

All authors are full-time employees and stockholders of Ibis Biosciences, Inc., a wholly owned subsidiary of Isis Pharmaceuticals, Inc.

ACKNOWLEDGMENTS

TIGER technology was jointly developed by the Ibis Biosciences a subsidiary of Isis Pharmaceuticals and SAIC (San Diego, USA). We acknowledge DARPA, CDC, Department of Homeland Security (DHS), and National Institute of Allergy and Infectious Diseases (NIAID) for financial support for development of the technology and Dr. Jacqueline R. Wyatt, Ph.D. (www.JL-SciEdit.com) for carefully reviewing the manuscript; patents are currently pending in the United States and internationally. Several key participants contributed significantly to the development and implementation of various aspects of the technology but are too numerous to be listed individually by name.

REFERENCES

1. Lam, S.K. 1998. Emerging infectious diseases–Southeast Asia. Emerg. Infect. Dis. **4:** 145–147.
2. Wong, K.T. *et al.* 2002. Nipah virus infection, an emerging paramyxoviral zoonosis. Springer Semin. Immunopathol. **24:** 215–228.
3. Wong, K.T. *et al.* 2002. Nipah virus infection: pathology and pathogenesis of an emerging paramyxoviral zoonosis. Am. J. Pathol. **161:** 2153–2167.
4. Cardosa, M.J. *et al.* 2003. Molecular epidemiology of human enterovirus 71 strains and recent outbreaks in the Asia-Pacific region: comparative analysis of the VP1 and VP4 genes. Emerg. Infect. Dis. **9:** 461–468.
5. Mackenzie, J.S. *et al.* 2001. Emerging viral diseases of Southeast Asia and the Western Pacific. Emerg. Infect. Dis. **7:** 497–504.
6. Poutanen, S.M. *et al.* 2003. Identification of severe acute respiratory syndrome in Canada. N. Engl. J. Med. **348:** 1995–2005.
7. Peiris, J. *et al.* 2003. Coronavirus as a possible cause of severe acute respiratory syndrome. Lancet **361:** 1319–1325.
8. Falsey, A.R. & E.E. Walsh. 2003. Novel coronavirus and severe acute respiratory syndrome. Lancet **361:** 1312–1313.
9. Ksiazek, T.G. *et al.* 2003. A novel coronavirus associated with severe acute respiratory syndrome. N. Engl. J. Med. **348:** 1953–1966.
10. Taylor, L.H., S.M. Latham & M.E. Woolhouse. 2001. Risk factors for human disease emergence. Philos. Trans. R. Soc. Lond. B Biol. Sci. **356:** 983–989.

11. SCHMIDT, T.M., E.F. DELONG & N.R. PACE. 1991. Analysis of a marine picoplankton community by 16S rRNA gene cloning and sequencing. J. Bact. **173:** 4371–4378.
12. KROES, I., P.W. LEPP & D.A. RELMAN. 1999. Bacterial diversity within the human subgingival crevice. Proc. Natl. Acad. Sci. (USA) **96:** 14547–14552.
13. STEPHENSEN, C.B., D.B. CASEBOLT & N.N. GANGOPADHYAY. 1999. Phylogenetic analysis of a highly conserved region of the polymerase gene from 11 coronaviruses and development of a consensus polymerase chain reaction assay. Vir. Res. **60:** 181–189.
14. OBERSTE, M.S., K. MAHER & M.A. PALLANSCH. 2002. Molecular phylogeny and proposed classification of the simian picornaviruses. J. Virol. **76:** 1244–1251.
15. MACK, D.H. & J.J. SNINSKY. 1988. A sensitive method for the identification of uncharacterized viruses related to known virus groups: hepadnavirus model system. Proc. Natl. Acad. Sci. USA **85:** 6977–6981.
16. ECHAVARRIA, M. *et al.* 1998. PCR method for detection of adenovirus in urine of healthy and human immunodeficiency virus-infected individuals. J. Clin. Micro. **36:** 3323–3326.
17. FOUT, G.S. *et al.* 2003. A multiplex reverse transcription-PCR method for detection of human enteric viruses in groundwater. Appl. Environ. Microbiol. **69:** 3158–3164.
18. BRITO, D.A., M. RAMIREZ & H. DE LENCASTRE. 2003. Serotyping *Streptococcus pneumoniae* by multiplex PCR. J. Clin. Microbiol. **41:** 2378–2384.
19. WILSON, K.H. *et al.* 2002. High-density microarray of small-subunit ribosomal DNA probes. Appl. Environ. Microbiol. **68:** 2535–2541.
20. WANG, D. *et al.* 2002. Microarray-based detection and genotyping of viral pathogens. Proc. Natl. Acad. Sci. USA **99:** 15687–15692.
21. WANG, D. *et al.* 2003. Viral discovery and sequence recovery using DNA microarrays. PLOS Biol. **1:** 257–260.
22. SAMPATH, R. & D.J. ECKER. 2004. Novel biosensor for infectious disease diagnostics. *In* Forum on Microbial Threats: Learning from SARS: Preparing for the Next Disease Outbreak—Workshop Summary. S.E. Knobler, A. Mahmoud & S. Lemon, Eds.: 181–185. The National Academies Press. Washington, DC.
23. SAMPATH, R. *et al.* 2005. Rapid identification of emerging pathogens: coronavirus. Emerg. Infect. Dis. **11:** 373–379.
24. ECKER, D.J. *et al.* 2005. Rapid identification and strain-typing of respiratory pathogens for epidemic surveillance. Proc. Natl. Acad. Sci. USA **102:** 8012–8017.
25. HOFSTADLER, S.A. *et al.* 2005. TIGER: the universal biosensor. Int. J. Mass Spectrom. **242:** 23–41.
26. VAN ERT, M. *et al.* 2004. Mass spectrometry provides accurate characterization of two genetic marker types in *Bacillus anthracis*. Biotechniques **37:** 642–651.
27. MACKE, T. *et al.* 2001. RNAMotif–An RNA secondary structure definition and search algorithm. Nuc. Acids Res. **29:** 4724–4735.
28. JIANG, Y. & S.A. HOFSTADLER. 2003. A highly efficient and automated method of purifying and desalting PCR products for analysis by electrospray ionization mass spectrometry. Anal. Biochem. **316:** 50–57.
29. MUDDIMAN, D.C. *et al.* 1997. Length and base composition of PCR-amplified nucleic acids using mass measurements from electrospray ionization mass spectrometry. Anal. Chem. **69:** 1543–1549.

30. SNIJDER, E.J. *et al.* 2003. Unique and conserved features of genome and proteome of SARS-coronavirus, an early split-off from the coronavirus group 2 lineage. J. Mol. Biol. **331:** 991–1004.
31. LINSSEN, B. *et al.* 2000. Development of reverse transcription-PCR assays specific for detection of equine encephalitis viruses. J. Clin. Microbiol. **38:** 1527–1535.
32. SIDWELL, R.W. & D.F. SMEE. 2003. Viruses of the *Bunya-Togaviridae* families: potential as bioterrorism agents and means of control. Antiviral Res. **57:** 101–111.
33. BRAULT, A.C. *et al.* 2004. Venezuelan equine encephalitis emergence: enhanced vector infection from a single amino acid substitution in the envelope glycoprotein. Proc. Natl. Acad. Sci. USA **101:** 11344–11349.
34. BRAULT, A.C. *et al.* 1999. Genetic and antigenic diversity among eastern equine encephalitis viruses from North, Central and South America. Am. J. Trop. Med. Hyg. **6:** 579–586.
35. ECKER, J.E. *et al.* 2006. High throughput genotyping of *Acinetobacter baumannii*. **44:** 2921–2932.

Bioproduction of Therapeutic Proteins in the 21st Century and the Role of Plants and Plant Cells as Production Platforms

ROBERT BOEHM

Institute of Plant Molecular Physiology and Biotechnology (IMBIO), University of Bonn, Bonn, Germany

ABSTRACT: In the last decade, the technique to genetically modify crop plants has gained more and more interest in terms of bioproduction of heterologous proteins. Plants have been discovered as a possible source for large amounts of cost effective recombinant protein. Main application fields are therapeutics for use in animal and human health, diagnostics, and technical enzymes. This review is focused on the recent progress in this field of molecular farming. After a comparison with hitherto established protein production systems, the advantages of plants as an alternative production system are discussed. An overview about the different host plants and possible expression strategies is given and the progress in commercialization of the techniques is highlighted. Finally, the role of plant cell cultures for the production of recombinant proteins is discussed.

KEYWORDS: bioproduction; gene expression; molecular farming; recombinant protein; transgenic plants

INTRODUCTION

Plants have been used for human benefit since the dawn of the human race, supplying the growing population with food, fibers, wood, and therapeutics. Thus, plants were, besides animals, the basic bioproduction system for important bulk substances. Especially in the field of pharmaceuticals, plants offer a huge variety of secondary metabolites with many therapeutic effects. These substances, evolutionarily developed by plants to protect themselves against pathogens and predators or to attract pollinators,[1] exhibit wound-healing, anti-inflammatory, antimicrobial, or psychoactive properties. They can be used

Address for correspondence: Robert Boehm, Institute of Plant Molecular Physiology and Biotechnology (IMBIO), Karlrobert Kreiten-Str. 13, 53115 Bonn, University of Bonn, Germany. Voice: +49-228-732156; fax +49-228-731696.
r.boehm@uni-bonn.de

Ann. N.Y. Acad. Sci. 1102: 121–134 (2007).
doi: 10.1196/annals.1408.009

to protect and maintain human and animal health. Modern biotechnological methods to produce these therapeutics in a controlled and predictable manner encompass cell cultures initiated from the respective medicinal plants. This sometimes is the only possibility for the large-scale production of these substances since the original plant species often are difficult to cultivate and are rare in nature. Successful examples for this kind of bioproduction are the alkaloids,[2] paclitaxel,[3] and shikonin.[4]

With the beginning of modern gene technology, it also became feasible to alter the plant genome by the addition or replacement of specific genes and their surrounding regulatory DNA sequences, for example, for tobacco[5] or sunflower.[6] This in the first instance was used to improve or alter the secondary products of therapeutic value that already were produced by plant cell cultures, for example, on tropan alkaloid or nikotin biosynthesis.[7,8] At the same time, first reports could demonstrate that plant cells are also able to express antibodies in a correct and biologically active way.[9,10] This opened up a completely new application field for transgenic plants. Here, the improvement of plant-derived substances or the plants themselves are no longer at the center of interest but the use of plants as a sole production tool for recombinant proteins of virtually any origin is of interest. These proteins can be purified after the plant harvest and applied for human benefit.

This review focuses on the current state of the use of intact plants and plant cell cultures as protein production systems for human and animal health in research and commercial applications. It highlights the potential that plants offer as a serious alternative to contribute to current protein production facilities in the future. Additionally, the use of plant cell cultures for this purpose is discussed.

ESTABLISHED PROTEIN PRODUCTION SYSTEMS

In the last decade, protein-based therapeutics have attracted more and more attention for the application in diagnosis and therapy. This includes mainly antibodies and antibody derivatives but also many serum-derived proteins (cytokines, growth hormones, interleukines, interferons, etc.). To date, 94 proteins of human therapeutic value have entered the market.[11] All these biologics are produced recombinantly with the state-of-the-art *in vitro* production systems: bacterial fermentation (*Escherichia coli*), yeast cell cultures (*Saccharomyces cerevisiae, Pichia pastoris*), or mammalian cell cultures. These production platforms, though very sophisticated and well characterized, still bear some disadvantages that are inherent in the respective production systems. Criteria for the assessment of the different systems are of biologic (yield of biologically active protein, posttranslational modifications, etc.) as well as economic nature (overall production costs, time to market, etc.).

To date, the bacterial fermentation is the best-characterized and most cost-effective production system. However, because of the procaryotic nature of the organisms, a eucaryotic type of protein processing is poorly developed and the production is limited to simple structures (peptides, small proteins) where posttranslational modifications are absent or unimportant for the biologic activity. Moreover, high protein accumulation often results in aggregation (inclusion bodies), thus, laborious renaturation steps are needed to recover the proteins in their biologically active form. The overall yield, however, remains relatively low, ranging from 100 to 1,500 mg/L.[12] Another point to consider is the bacterial toxins that have to be eliminated during downstream processing. Yeasts as eucaryotic organisms avoid some of these disadvantages, but the practical experiences show a high amount of product loss due to degradation of the target protein in the medium. Moreover, especially *S. cerevisiae* tends to hyperglycosylate the produced proteins what results in N-glycans of high-mannose structure, which are totally different from the mammalian or human structures of N-glycans. These critical points can be improved by using *P. pastoris* as expression system.[13] Here, a protein yield up to 6.4 g/L could be obtained,[14] although the average protein yield is 100–200 mg/L.[13] Despite these limitations, 43% of the current biologics on the market are produced in bacteria and yeasts. The majority of the biologics (57%), however, is produced in different mammalian cell cultures (e.g., NS0, BHK, CHO). As cell types most similar to their human counterpart, these systems offer the highest yield of functional recombinant protein with correct N-glycosylation and other posttranslational modifications. The yield typically ranges between 1 and 3 g/L.[12] This "golden standard," however, requires expensive infrastructure and media components. Additionally, the risk of viral or oncogenic contamination is given. This leads to reasonably high overall production costs, ranging from 300 to 10,000 USD/g, depending of the quantity of the protein produced.[15]

FUTURE QUANTITY DEMAND OF THERAPEUTIC PROTEINS

The human genome project has also boosted the identification of target proteins of therapeutic potential. Thus, the pipeline of protein-based therapeutics in preclinic and clinical trials at the moment by far exceeds the amount of such kind of therapeutics on the market.[11] Within this target pipeline, the portion of monoclonal antibodies will enlarge from today's 25% of all biologics to nearly 62%, mainly based in the therapeutic field of cancer treatment, inflammatory diseases, and anti-infective therapy.[11] After these drug candidates have been approved for the market, the quantity demand for each will range from hundreds of kilograms to several metric tons of highly pure protein per year. This demand exceeds today's industrial production capacities and even after increase of manufacturing capacities, this issue is regarded as a future bottleneck in bringing protein-based therapeutics to the market.[16] Even when this capacity problem is solved using the established production systems, the high investment and production costs will put reasonable financing pressure on the national health systems. From that point of view, new technologies for large-scale and cost-effective biomanufacturing of therapeutic proteins are highly appreciated.

WHY PLANTS AS PRODUCTION PLATFORM?

Since the early days of expression of antibodies in plants,[9,10] many feasibility studies of the production of different therapeutic proteins have been performed in plants[17,18] and plant cell cultures.[19,20] This includes IgG,[21] acetylcholinesterase,[22] α-interferon,[23] human α-1-antitrypsin,[24] or secretory IgA,[25] which cannot be produced in any other production system so far. Plant cells as eucaryotic systems possess all the features for the biologically active generation of complex therapeutic proteins. This results in a very high ratio of biologically active recombinant protein. In a comparative study of the expression of anti-CEA scFv, tobacco BY-2 cells showed by far the highest amount of functional protein (92%) of total recombinant protein (TP), compared to *E. coli* (12% TP) and *P. pastoris* (40% TP).[12]

Using open-field cultivation, transgenic plants have superior advantages over all established expression systems in terms of capacity, flexibility, scalability, and production costs. If transgenic seed material can be supplied to the market in sufficient quantities, farming will rely on the well-established, low-cost agricultural infrastructure. The acreage of planted transgenic plants can be adopted to meet altered quantity demands from one year to another without additional costs. Taking into account an average protein yield of 0.5–1 g per kg fresh weight, Fischer and coworkers[26] calculate that 1 hectare of tobacco plants would result in 100 tons of fresh material per year. From this, 50 kg of sIgA or 100 kg of recombinant glucocerebrosidase can be harvested. Thus, the relatively low protein yield, which is a general feature of transgenic plants compared to other expression systems, can be compensated easily by the huge quantity of generated biomass. Moreover, the downstream processing costs may also be diminished because there is no need to downgrade any bacterial toxins or human pathogenic particles (viruses, toxins, prions) existing in animal production systems. Taking these unique features into account, the costs for the resulting proteins have been calculated to be below 50 USD per gram.[27]

CHOOSING A HOST PLANT FOR THE GENE EXPRESSION

For the production of valuable secondary metabolites, one is restricted to the plant species where this specific and complex biosynthetic pathway can be found and exploited. This entire pathway cannot be transferred to another plant. In contrast, proteins are encoded by only one or very few genes that can be inserted and expressed in virtually any host plant. This gives much more freedom to operate, but, on the other hand, leads to a vast variety of possible host plants and expression strategies for the production of recombinant proteins. There is not one clear advantageous, leading plant expression system, although the well-established crop species are preferably used, including tobacco, rice, and corn.[15,18] Depending on the biologic and economic requirements of one specific target protein, one is free to choose the most suitable plant expression system. There are two basic strategies for the expression: transient expression or stably transformed plants. Transient expression has some advantages since this technique does not result in transgenic plants with its ecologic or regulatory concerns. Moreover, a reasonable amount of protein can be obtained within few days after infection.[28] The transient expression can be achieved using infiltration of whole plants or plant parts (e.g., leaves) with *Agrobacterium tumefaciens*.[29,30] This results in lab-scale quantities of protein in the milligram range for feasibility or preclinical studies. In another transient expression strategy, the target gene is fused to the coat protein gene of plant viruses, for example, tobacco mosaic virus, and the thus altered virus is used for plant infection.[28] This technique is used commercially for large-scale protein production in open-field plants, for example, by Large Scale Biology's Geneware® technology[31] and Icon Genetics's MagICON® technology.[32] Transgenic plants, on the other hand, require a longer time for the generation and characterization of production lines and the generation of a sufficient amount of seed material. But after this, an almost unlimited and sustainable production capacity can be reached using the established agricultural infrastructure. This strategy is already applied commercially using bulk crop plants like tobacco, corn, rice, or soybean[15,18] or special plant species like soapworts (Saponin Inc., Saskatoon, Canada), safflower (SemBioSys Genetics Inc., Calgary, Alberta Canada), or alfalfa (Medicago Inc., Quebec Canada). An alternative to open field plants is the cultivation of the transgenic plants in greenhouses, which enables more controlled production conditions, but increases the production costs. This decision depends on the quantity demand of the target protein and the regulatory issues of the specific country in terms of release of transgenic plants into the environment and GMP requirements of the production process.

AQUATIC PLANTS AS EXPRESSION SYSTEMS

Free floating water plants as production hosts for recombinant protein production are a quite new and innovative field of molecular farming, providing a further alternative to open-field plants. This includes higher, free floating water plants as well as algae. All these have the advantage of a fast vegetative growth rate and the cost-effective cultivation on water and inorganic nutrients. Depending on the safety and mass requirements, these organisms can be held in contained *in vitro* cultures, in controlled greenhouse containers, or in open ponds of any size for bulk biomass production. From eucaryotic algae, mostly

Chlamydomonas reinhardtii is used.[33] But other algae species eventually have been tested for that purpose, for example, *Chlorella ellipsoidea*.[34] The specific algae-related problem of a codon usage different from animal and human could be overcome by the use of completely synthetic genes with adopted codon usage.[35]

Regarding higher plants, members of the *Lemnaceae* (duckweed) family successfully are already commercially exploited as a source for recombinant therapeutic proteins, namely *Lemna minor* by Biolex Inc. (Pittsboro, NC,USA) and *Spirodela oligorhiza* by LemnaGene (France). Also, species of the genus *Wolffia* are under development for that purpose.[36,37] A specific advantage of duckweed is that these plants are rich in proteins and can serve as animal feed.[38] In addition, *Wolffia* plants were also cultivated as "water eggs" and used for human nutrition in east Asia in former times.[39]

SPATIAL AND TEMPORAL REGULATION OF GENE EXPRESSION

In most cases, the target protein is expressed in the whole plant and the gene is driven by constitutive promoters, for example, CaMV 35S or *mas* promoter without any regulation. This is to reach the maximum protein yield. But the temporal and spatial regulation of transgene expression within the plant can be advantageous for specific applications. This regulation is reached by the use of plant promoters that are active only in a specific plant organ or time frame within the plant's life. If the transgene is expressed constitutively in the whole plant, the recombinant protein has to be extracted right after the plant harvest and the crude protein extract or the purified protein has to be distributed under elaborate conditions in a cooling chain. In contrast, targeting the expression to one of the plant storage organs leads to an accumulation of the resulting protein exclusively in these plant parts. Here, the protein remains stable and active at ambient temperature for an extended time span. After harvest, the organs can be stored and transported without any cooling or stabilization and can be extracted at a place distant from the harvesting location. The target storage organs mostly were seeds (cereals, corn) or tubers (potato). This strategy is applied successfully to several technical proteins and enzymes and is also considered for therapeutic protein production.[40] To make protein production in plants more safe and controllable, it can be useful to put the target gene under the control of an inducible promoter. This avoids transgenic plants containing recombinant protein on the field and the deliberate induction of gene expression enables protein production after the harvest right before extraction. This possibility successfully has been applied using a stress-inducible promoter (MeGA^{TM} promoter) in tobacco by CropTech Corp (Blacksburg, VA, USA).[41] Here, the foreign gene expression is induced by chopping the freshly harvested tobacco plants under contained conditions. In another strategy, a germination-specific promoter drives gene expression in barley seeds. After harvest and storage of the seed granules, the protein production can be induced at the desired time point and place by germinating (malting) the granules. This technique is proprietary of MALTagen GmbH (Andernach, Germany).

EFFECTS ON THE INTRACELLULAR TARGETING OF THE RECOMBINANT PROTEIN

The intracellular targeting of the resulting protein is of great influence to protein yield and its biological properties. At the current state of the art, the targeting of all intracellular compartments is feasible using known signal sequences that can be fused to the target gene. The intracellular locations that have been tested and compared for foreign protein production include cytosol, the endoplasmic reticulum (ER), the apoplastic space, the vacuole, and the chloroplast (see below). Several studies have been performed to compare the efficacy of protein production using cytoplasmic, ER, and apoplastic accumulation in tobacco[42–44] or rice.[45] Conclusive, ER targeting gives rise to the highest yield of biologically active protein. In contrast, cytosolic expression mostly leads only to poor protein yield. The reason is that the ER contains many important posttranslational protein maturation elements like chaperones, protein-disulfide isomerase, and glycosylation enzymes in a reducing environment. These features are not at all or scarcely present in the cytosol. Vacuolar targeting and secretion into the apoplastic space is also coupled with ER/Golgi passage and the above-mentioned maturation and stabilization properties. However, the protein yield often is found to be diminished compared to ER accumulation, probably due to unfavorable protein stability properties of these compartments.

CHLOROPLAST TRANSFORMATION

The targeting of the foreign gene's expression to the chloroplasts is a completely different strategy that cannot be accomplished using the respective signal sequence for chloroplast protein import but by a specific transformation method of the chloroplast itself using particle bombardment.[46,47] Successful chloroplast transformation, which is selected via chloroplast-specific selection markers, leads to "transplastomic" rather than transgenic plants. After the generation of fully homotransplastomic plants by repeated rounds of selection, the protein yield reaches up to 46.1% total leaf protein[46] and, thus, is superior to all other transgenic expression strategies with regard to protein yield. This results from the high copy number of the chloroplast genome and of the integrated transgene coupled with high transcription rates. Some more advantages render the chloroplastic expression of foreign genes to a promising strategy to compete with other nonplant expression systems regarding protein yield: Due to the procaryotic nature of the chloroplasts, it is possible to transfer whole polycistrons with several genes under the same regulation and to get their coordinate expression. Moreover, homologous recombination is feasible to integrate genes into specific genome sites without gene silencing or position effects, which leads to reduced variability in expression levels of the introduced transgene. Finally, the plants lack transgenic pollen due to the maternal inheritance of the chloroplast genome. The only disadvantage that results from the procaryotic organelle character is the decreased ability for necessary posttranslational modifications present in the ER. Thus, the range of proteins that can be obtained in a biologically active manner from chloroplast transformation is limited similarly to the bacterial expression systems. The chloroplast transformation was first established in tobacco, but the range of transformable plant species is increasing in recent years, including cotton,[48] soybean,[49] oilseed rape,[50] tomato,[51] and potato.[52] A growing number of therapeutic relevant proteins have been expressed using this strategy, including interferon gamma[53] vaccines[54–56] and other therapeutic relevant proteins.[46] This production strategy has already been commercialized, for example, by Chlorogen, Inc. (St. Louis, MO, USA) and Icon Genetics (Halle, Germany).

EDIBLE VACCINES

Besides living, attenuated, and inactivated strains of viruses and bacteria, it has been found that also isolated parts or even single proteins of these pathogens (so called subunit vaccines) can provoke an active, prophylactic immunization, if delivered by injection. These subunit vaccines are another class of therapeutic proteins that can actively be expressed using plants.[57] An alternative vaccine delivery to reach an active immunization efficiently is oral or nasal exposure of the vaccines to mucosal tissue of nose, throat, and gut. Here, the vaccines elicit the mucosal (sIgA) as well as systemic (IgG) immune response.[58] These findings paved the road to a very fascinating combination of medicine and plant science: The plants expressing oral vaccines do not only serve as the production system but at the same time as delivery vehicle for the vaccines. Oral immunization simply can be achieved by eating edible plant parts expressing the subunit vaccine gene. This enormously simplifies delivery and administration and reduces the production costs, since any kind of downstream processing and protein purification is unnecessary. Thus, the production of vaccines in edible plant parts like fruits is of unique advantage. Choosing an appropriate host species like banana, it opens up the possibility for low-cost immunization in underprivileged countries with poor medical supply.

Vaccines belonged to the first recombinant proteins that have been generated using transgenic plants.[59] The successful active immunization after oral administering of vaccine-expressing plant parts has been proven several times.[60]

Since then, numerous vaccines have been produced in transgenic plants, including hepatitis B envelope surface protein, Norwalk virus capsid protein, heat-labile toxin B-subunit of enterotoxigenic *E. coli* for human immunization, and VP1 gene of foot-and-mouth disease virus or glycoprotein S of transmissible gastroenteritis coronavirus for veterinary applications.[15] Relevant host plants for the expression were especially those generating edible plant parts that can be eaten without any processing or cooking. This includes tomato[61] and carrot.[62] Although no report about vaccine production in banana is available to date, this plant species is especially prone for vaccine production in developing countries, where it can be cultivated easily. Additionally, leaves can in some cases be eaten raw and, thus, used for oral vaccine production, as for example, alfalfa[63] or lettuce.[64] Potato tubers as an accumulation organ for vaccines often have successfully been used as model system[65] but lack practical relevance, since potatoes have to be cooked for human food, destroying much of the vaccine protein. In conclusion, if some regulatory issues like homogenous vaccine dose per gram fresh weight can be solved satisfactorily, plant-derived vaccines will hold many promises for safe and cost-effective mass immunization of mankind much more than any other production system.

COMMERCIALIZATION OF PLANT-BASED PROTEIN PRODUCTION

Because the perspective for an economically reasonable protein production in transgenic plants is that promising, several companies, especially in the United States and Canada, have been founded that focus on the production of therapeutic proteins in open-field-cultivated plants, including Large Scale Biology (www.lsbc.com), Ventria Bioscience (www.ventria.com), Planet Biotechnology (www.planetbiotechnology.com), Dow Agroscience (www.dowagro.com), and Agracetus (www.agracetus.com) in the United States, Medicago (www.medicago.com), Plantigen (www.lhsc.on.ca/plantigen), and SemBioSys (www.sembiosys.ca) in Canada, Icon Genetics (www.icongenetics.com), Novoplant (www.novoplant.de), Sungene (www.sungene.de), and MALTagen (www.maltagen.de) in Germany, or Méristem Therapeutics (www.meristem-therapeutics.com) in France. Many of the numerous therapeutic proteins under development have already entered clinical phase trials. However, no plant-derived product has entered the market to date. The first product, a sIgA/IgG chimeric antibody against *Streptococcus mutans* for caries treatment, will be launched by Planet Biotechnology in 2006, according to the company.[66]

PLANT *IN VITRO* CULTURES

In contrast, production systems using plant *in vitro* cultures are not used at a commercial level so far, but several systems are under development at the academia.[19,20,67] Plant cell cultures are more comparable to mammalian cell cultures from the view of biochemical engineering. Although no calculation of overall production costs using plant cell cultures can be found in literature, these systems promise to be more cost effective. This is due to the simple media using cheaper medium components (mostly inorganic salts and sucrose) and the lack of human pathogenic particles that have to be eliminated in the downstream processing. As a first estimation, plant cell cultures cannot compete with transgenic plants in open-field cultivation with regard to production costs, but still promise to be more cost-effective than mammalian cell cultures. This, however, is also strictly dependent on the level of gene expression and, thus, protein yield. First feasibility studies demonstrate the successful expression of IgG,[68] interleukin-12,[69] human granulocyte-macrophage colony stimulation factor,[70] or immunotoxins.[71] However, protein yield remains low, averaging 1–5 mg/L cell suspension culture.[19] Exceptional high protein yield has been reported rarely, for example, 200 mg/L of α-1-antitrypsin in rice cell cultures.[72] This, however, highlights the potential of plant *in vitro* systems for therapeutic bioproduction purposes. To finally reach economical profitability, much effort has to be put on consequent strain and genotype selection, media component optimization, development of appropriate promoters and enhancer elements, codon optimization, et cetera. to enhance productivity by a factor of at least 500. This has been done in a similar way with animal cell culture strains over the last 20 years, which finally led to today's productivity in this area. It remains questionable, however, whether there is enough interest from the industry side to push the development of such profitable plant cell culture systems in the future in the face of available, optimized, and approved animal cell production lines. If so, the development will certainly focus on few well-characterized lines, for example, BY-2 tobacco cells,[73] rice cell culture,[45] or that of the moss *Physcomitrella patens*.[74]

REFERENCES

1. WINK, M. 1988. Plant breeding – Importance of plant secondary metabolites for protection against pathogens and herbivores. Theor. Appl. Genet. **75:** 225–233.
2. VERPOORTE, R., A. CONTIN & J. MEMELINK. 2002. Biotechnology for the production of plant secondary metabolites. Phytochem. Rev. **1:** 13–25.
3. ZONG, J.J. 2002. Plant cell culture for production of paclitaxel and other taxanes. J. Biosci. Bioeng. **94:** 591–599.
4. YAZAKI, K., H. MATSUOKA, T. UJIHARA & J. SCHELL. 1999. Shikonin biosynthesis in *Lithospermum erythrorhizon*: light-induced negative regulation of secondary metabolism. Plant Biotechnol. **16:** 335–342.
5. HORSCH, R.B., J.E. FRY, N.L. HOFFMANN, *et al.* 1985. A simple and general method for transferring genes into plants. Science **227:** 1229–1231.
6. EVERETT, N.P., K.E.P. ROBINSON & D. MASCARENHAS. 1987. Genetic engineering of sunflower (*Helianthus annuus* L.). Bio/Technology **5:** 1201–1294.

7. YUN, D., T. HASHIMOTO & Y. YAMADA. 1992. Metabolic engineering of medicinal plants: transgenic *Atropa belladonna* with an improved alkaloid composition. Proc. Natl. Acad. Sci. USA **89:** 11799–11803.
8. HAMILL, J.D., R.J. ROBINS, A.J. PARR, *et al.* 1990. Over-expressing a yeast ornithine decarboxylase gene in transgenic roots of *Nicotiana rustica* can lead to enhanced nicotine accumulation. Plant Mol. Biol. **15:** 27–38.
9. HIATT, A., R. CAFFERKEY & K. BOWDIS. 1989. Production of antibodies in transgenic plants. Nature **342:** 76–77.
10. DÜRING, K., S. HIPPE, F. KREUZALER & J. SCHELL. 1990. Synthesis and self-assembly of a functional mononclonal antibody in transgenic *Nicotiana tabacum*. Plant Mol. Biol. **15:** 281–293.
11. STEINER, U. 2005. State of the biologic market. Oral presentation held at the Conference of plant-made pharmaceuticals, Montréal (http://archives.cpmp2005.org).
12. FISCHER, R. 2005. Qualitative and quantitative comparison of protein expression systems. Oral presentation held at the Conference of plant-made pharmaceuticals, Montréal (http://archives.cpmp2005.org).
13. MACAULEY-PATRICK, S., M.L. FAZENDA, B. MCNEIL, & L.M. HARVEY. 2005. Heterologous protein production using the *Pichia pastoris* expression system. Yeast **22:** 249–270.
14. CHEN, C.-C., P.-H. WU, C.-T. HUANG & K.-J. CHENG. 2004. A *Pichia pastoris* fermentation strategy for enhancing the heterologous expression of a *Escherichia coli* phytase. Enz. Microb. Technol. **35:** 315–320.
15. DANIELL, H., S.J. STREATFIELD & K. WYCOFF. 2001. Medical molecular farming: production of antibodies, biopharmaceuticals and edible vaccines on plants. Trends Plant Sci. **6:** 219–226.
16. DEPALMA, A. 2002. Capacity mismatch for production of monoclonals. Gen. Eng. News **22:** 62–66.
17. MA, J.K.-C., P.M.W. DRAKE & P. CHRISTOU. 2003. The production of recombinant pharmaceutical proteins in plants. Nat. Rev. Genet. **4:** 794–805.
18. GIDDINGS, G., G. ALLISON, D. BROOKS & A. CARTER. 2000. Transgenic plants as factories for biopharmaceuticals. Nat. Biotech. **18:** 1151–1155.
19. HELLWIG, S., J. DROSSARD, R.M. TWYMAN & R.FISCHER. 2004. Plant cell cultures for the production of recombinant proteins. Nat. Biotech. **22:** 1415–1422.
20. DORAN, P.M. 2000. Foreign protein production in plant tissue culture. Curr. Opin. Biotech. **11:** 199–204.
21. STÖGER, E., M. SACK, R. FISCHER & P. CHRISTOU. 2002a. Plantibodies: application, advantages and bottlenecks. Curr. Opin. Biotechnol. **13:** 161–166.
22. MOR, T.S., M. STERNFELD, H. SOREQ, *et al.* 2001. Expression of recombinant human acetylcholinesterase in transgenic tomato plants. Biotechnol. Bioeng. **75:** 259–266.
23. SAWAHEL, W.A. 2002. The production of transgenic potato plants expressing human alpha-interferon using lipofectin-mediated transformation. Cell Mol. Biol. Lett. **7:** 19–29.
24. TREXLER, M.M., K.A. MCDONALD & A.P. JACKMAN. 2005. A cyclical semicontinuous process for production of human alpha 1-antitrypsin using metabolically induced plant cell suspension cultures. Biotechnol. Prog. **21:** 321–328.
25. MA, J.K.-C, B.Y. HIKMAT, K. WYCOFF, *et al.* 1998. Characterization of a recombinant plant monoclonal secretory antibody and preventive immunotherapy in humans. Nat. Med. **4:** 601–606.

26. Fischer, R., S. Schillberg & N. Emans. 2001. Molecular farming: a field of growing promise. Outlook Agr. **30:** 31–36.
27. Evangelista, R.L., A.R. Kusnadi, J.A. Howard & Z. L.Nikolov. 1998. Process and economic evaluation of the extraction and purification of recombinant b-glucuronidase from transgenic corn. Biotechnol. Prog. **14:** 607–614.
28. Fischer, R., C. Vaquero-Martin, M. Sack, *et al.* 1999b. Towards molecular farming in the future: transient protein expression in plants. Biotechnol. Appl. Biochem. **30:** 113–114.
29. Vaquero, C., M. Sack, J. Chandler, *et al.* 1999. Transient expression of a tumor-specific single-chain fragment and a chimeric antibody in tobacco leaves. Proc. Natl. Acad. Sci. USA **96:** 11128–11133.
30. Kapila, J., R. de Rycke, M. van Montagum & G. Angenon. 1997. An *Agrobacterium*-mediated transient gene expression system for intact leaves. Plant Sci. **122:** 101–108.
31. Pogue, G.P., J.A. Lindbo, W.O. Dawson & T. H. Turpen. 1998. Tobamovirus transient expression vectors: tools for plant biology and high-level expression of foreign proteins in plants. *In* Plant Molecular Biology Manual. S.B. Gelvin & R.A. Schilperoot, Eds.: L4, 1–27. Kluwer Academic Publishers. Dordrecht, the Netherlands.
32. Marillonnet, S., C. Thoeringer, R. Kandzia, *et al.* 2005. Systemic *Agrobacterium tumefaciens*-mediated transfection of viral replicons for efficient transient expression in plants. Nat. Biotech. **23:** 718–723.
33. Franklin, S.E. & S.P. Mayfield. 2005. Recent developments in the production of human therapeutic proteins in eucaryotic algae. Expert Opin. Biol. Th. **5:** 225–235.
34. Kim, D.H., Y.T. Kim, J.J. Cho, *et al.* 2002. Stable integration and functional expression of flounder growth hormone gene in transformed microalga, *Chlorella ellipsoidea*. Mar. Biotechnol. **4:** 63–73.
35. Fuhrmann, M., W. Oertel & P. Hegemann. 1999. A synthetic gene coding for the green fluorescent protein (GFP) is a versatile reporter in *Chlamydomonas reinhardtii*. Plant J. **19:** 353–361.
36. Boehm, R., C. Kruse, D. Voeste, *et al.* 2001. A transient transformation system for duckweed (*Wolffia columbiana*) using *Agrobacterium*-mediated gene transfer. J. Appl. Bot. **75:** 107–111.
37. Kruse, C., R. Boehm, D. Voeste, *et al.* 2001. Transient transformation of *Wolffia columbiana* by particle bombardment. Aquatic Bot. **72:** 175–181.
38. Islam, K.M.S. 2002. Feasibility of duckweed as poultry feed—a review. Indian J. Anim. Sci. **72:** 486–491.
39. Appenroth, K.-J. & H. Augsten. 1996. Wasserlinsen und ihre Nutzung. Biologie in unserer Zeit **26:** 187–194.
40. Stöger, E., J.K.C. Ma, R. Fischer & P. Christou. 2005. Sowing the seeds of success: pharmaceutical proteins from plants. Curr. Opin. Biotech. **16:** 167–173.
41. Cramer, C.L. & D.L. Weissenborn. 1997. HMG2 promoter expression system and post-harvest production of gene products in plants and plant cell cultures. US. patent 5,670,349.
42. Schillberg, S., S. Zimmermann, A. Voss & R. Fischer. 1999. Apoplastic and cytosolic expression of full-size antibodies and antibody fragments in *Nicotianum tabacum*. Transgenic Res. **8:** 255–263.

43. FISCHER, R., D. SCHUMANN, S. ZIMMERMANN, *et al.* 1999c. Expression and characterization of bispecific single-chain Fv fragments produced in transgenic tobacco. Eur. J. Biochem. **262:** 810–816.
44. STÖGER, E., M. SACK, Y. PERRIN, *et al.* 2002b. Practical consideration for pharmaceutical antibody production in different crop systems. Mol. Breeding **9:** 149–158.
45. TORRES, E., C. VAQUERO, L. NICHOLSON, *et al.* 1999. Rice cell culture as an alternative production system for functional diagnostic and therapeutic antibodies. Transgenic Res. **8:** 441–449.
46. DANIELL, H., S. CHEBOLU, S. KUMAR, *et al.* 2005. Chloroplast-derived vaccine antigens and other therapeutic proteins. Vaccine **23:** 1779–1783.
47. BOCK, R.. 2001. Transgenic plastids in basic research and plant biotechnology. J. Mol. Biol. **312:** 425–438.
48. KUMAR, S., A. DHINGRA & H. DANIELL. 2004. Stable transformation of the cotton plastid genome and maternal inheritance of transgenes. Plant Mol. Biol. **56:** 203–216.
49. DUFOURMANTEL, N., B. PELISSIER, F. GARCON, *et al.* 2004. Generation of fertile transplastomic soybean. Plant Mol. Biol. **55:** 479–489.
50. HOU, B.K., Y.H. ZHOU, L.H. WAN, *et al.* 2003. Chloroplast transformation in oilseed rape. Transgenic Res. **12:** 111–114.
51. RUF, S., M. HERMANN, I.J. BERGER, *et al.* 2001. Stable genetic transformation of tomato plastids and expression of a foreign protein in fruit. Nat. Biotech. **19:** 870–875.
52. SIDOROV, V.A., D. KASTEN, S.Z. PANG, *et al.* 1999. Stable chloroplast transformation in potato: use of green fluorescent protein as a plastid marker. Plant J. **19:** 209–216.
53. LEELAVATHI, S. & V.S. REDDY. 2003. Chloroplast expression of His-tagged GUS-fusions: a general strategy to overproduce and purify foreign proteins using transplastomic plants as bioreactors. Mol. Breeding **11:** 49–58.
54. KANG, T.J., S.C. HAN, M.Y. KIM, *et al.* 2004. Expression of non-toxic mutant of *Escherichia coli* heat-labile enterotoxin in tobacco chloroplasts. Protein Exp. Purif. **38:** 123–128.
55. TREGONING, J., P. MALIGA, G. DOUGAN & P. J. NIXON. 2004 New advances in the production of edible plant vaccines: chloroplast expression of a tetanus vaccine antigen, TetC. Phytochem. **65:** 989–994.
56. SUN, M., K.X. QIAN, N. SU, *et al.* 2003. Foot-and-mouth disease virus VP1 protein fused with cholera toxin B subunit expressed in *Chlamydomonas reinhardtii* chloroplast. Biotechnol. Lett. **25:** 1087–1092.
57. STREATFIELD, S.J. & J.A. HOWARD. 2003. Plant-based vaccines. Int. J. Parasitol. **33:** 479–493.
58. WARZECHA, H. 2000. Essbare Impfstoffe. DAZ **31:** 45–50.
59. MASON, H.S., D.M. LAM & C.J. ARNTZEN. 1992. Expression of hepatitis B surface antigen in transgenic plants. Proc. Natl. Acad. Sci. USA **96:** 11128–11133.
60. WALMSLEY, A.M. & C.J. ARNTZEN. 2003. Plant cell factories and mucosal vaccines. Curr. Opin. Biotech. **14:** 145–150.
61. HUANG, Z., G. ELKIN, B.J. MALONEY, *et al.* 2005. Virus-like particle expression and assembly in plants: hepatitis B and Norwalk viruses. Vaccine **23:** 1851–1858.
62. IMANI, J., A. BERTING, S. NITSCHE, *et al.* 2002. The integration of a major hepatitis B virus gene into cell-cycle synchronized carrot cell suspension cultures and its expression in regenerated carrot plants. Plant Cell Tiss. Org. Cult. **71:** 157–164.

63. WIGDOROVITZ, A., D.M.P. FILGUEIRA, N. ROBERTSON, *et al.* 1999. Protection of mice against challenge with foot and mouth disease virus (FMDV) by immunization with foliar extracts from plants infected with recombinant tobacco mosaic virus expressing the FMDV structural protein VP1. Virology **264:** 85–91.
64. KAPUSTA, J., A. MODELSKA, M. FIGLEROWICZ, *et al.* 1999. A plant-derived edible vaccine against hepatitis B virus. FASEB J. **13:** 2339–2340.
65. KONG, Q., L. RICHTER, Y.Y. FANG, *et al.* 2001. Oral immunization with hepatitis B surface antigen expressed in transgenic plants. Proc. Natl. Acad. Sci. USA **98:** 11539–11544.
66. FINEMAN, E. 2005. Host presentation of Planet Biotechnology. Oral presentation held at the Conference of plant-made pharmaceuticals, Montréal (http://archives.cpmp2005.org).
67. FISCHER, R., N. EMANS, F. SCHUSTER, *et al.* 1999a. Towards molecular farming in the future: using plant-cell-suspension cultures as bioreactors. Biotechnol. Appl. Biochem. **30:** 109–112.
68. SHARP, J.M. & P.M. DORAN. 2000. Characterization of monoclonal antibody fragments produced by plant cells. Biotechnol. Bioeng. **73:** 338–346.
69. KWON, T.H., J.E. SEO, J. KIM, *et al.* 2003a. Expression and secretion of the heterodimeric protein Interleukin-12 in plant cell suspension culture. Biotechnol. Bioeng. **81:** 870–875.
70. KWON, T.-H., Y.-S. KIM, J.-H. LEE & M.-S.YANG. 2003b. Production and secretion of biologically active human granulocyte-macrophage colony stimulation factor in transgenic tomato suspension cultures. Biotechnol. Lett. **25:** 1571–1574.
71. FRANCISCO, J.A., S.L. GAWLAK, M. MILLER, *et al.* 1997. Expression and characterization of Bryodin 1 and a Bryodin 1-based single-chain immunotoxin from tobacco cell culture. Bioconjugate Chem. **8:** 708–713.
72. HUANG, J.M., T.D. SUTLIFF, L.Y. WU, *et al.* 2001. Expression and purification of functional human alpha-1-antitrypsin from cultured plant cells. Biotechnol. Prog. **17:** 126–133.
73. NAGATA, T., Y. NEMOTO & S. HASEZAWA. 1992. Tobacco BY-2 cell line as the "HeLa" cell in the cell biology of higher plants. Int. Rev. Cytol. **132:** 1–30.
74. DECKER, E.L. & R. RESKI. 2004. The moss bioreactor. Curr. Opin. Plant Biol. **7:** 166–170.

West Nile Virus

Strategies for Predicting Municipal-Level Infection

NIKOLAOS W. YIANNAKOULIAS[a] AND LAWRENCE W. SVENSON[a,b,c]

[a]*Public Health Surveillance and Environmental Health, Alberta Health and Wellness, Edmonton, Alberta, Canada*

[b]*Department of Public Health Sciences, University of Alberta, Edmonton, Alberta, Canada*

[c]*Department of Community Health Sciences, University of Calgary, Calgary, Alberta, Canada*

ABSTRACT: The appearance and spread of West Nile virus (WNv) in North America represent a recent example of how mosquito-borne diseases can develop in new settings. Understanding the epidemiological, biological, and geographical aspects of WNv is critical to developing a greater understanding of how newly emerging, migrating, or evolving vector-borne infectious disease can develop globally. To aid in the allocation of resources that mitigate future outbreaks and to better understand the geographic nature of WNv in the North American prairies, we employ spatial and nonspatial modeling methods to predict municipal-level risk of human WNv infection rates. We use data based on a combination of routinely collected electronic data sources. Our findings suggest general agreement between spatial and nonspatial approaches, and results are consistent with seroprevalence-based estimates. We suggest that spatial models based on administrative data can offer estimates of relative risk in human populations at less cost, and in a timelier manner than estimates based on serology specimens.

KEYWORDS: *Cx. tarsalis*; geographic information systems (GIS); seroprevalence; West Nile virus (WNv)

INTRODUCTION

West Nile Virus in North America

The first appearance of West Nile virus (WNv) in North America was the 1999 New York City outbreak.[1] Since then, the virus has spread

Address for correspondence: Lawrence W. Svenson, Health Surveillance, Alberta Health and Wellness, P.O. Box 1360 Stn. Main, Edmonton, Alberta, Canada T5J 2N3. Voice: +780-422-4767; fax: 780-427-1470.
larry.svenson@gov.ab.ca

Ann. N.Y. Acad. Sci. 1102: 135–148 (2007).
doi: 10.1196/annals.1408.010

westward, reaching the western parts of North America, Central America, and the Caribbean, and causing local outbreaks across the continent. In 2003, the Canadian prairies experienced its first active year of WNv infection, including the positive identification of 1,365 human cases (Public Health Agency of Canada, http://www.phac-aspc.gc.ca/wnv-vwn/mon-hmnsurv-archive˙e.html). Most persons infected with WNv are asymptomatic,[2] and the total infection rate in the population was probably higher than reported, with seroprevalence estimates as high as 5% to 15% of the population in some regions.[3,4] In a small percentage of cases, symptoms are more severe, resulting in meningitis, encephalitis, and other neurological illness.[5] Although the likelihood of infection does not appear to be age specific, severe neurological symptoms are more common in those over 60 years of age.[6]

Most human infection occurs as a result of bites from mosquitoes carrying the virus, though WNv persists in an enzootic cycle involving birds and mosquitoes. In North America, it is believed that most bird–mosquito viral amplification occurs by the late summer.[7] Mosquito species with low laboratory vector competence may not play an important role in the amplification of the virus, but may be responsible for transferring WNv to humans and other mammals.[8] Several North American mosquito species may be involved in the WNv transmission cycle and in particular several *Culex* varieties.[5,8–10] Of these, *Culex tarsalis* is perhaps the most likely vector in western North America, given its abundance in the southern non-Alpine regions of the North American prairies.[11] Several other prairie-dwelling mosquito species are competent vectors of WNv, however, *Cx. tarsalis* feeds on both mammalian and avian hosts, particularly in the summer and fall months. These feeding habits may make it well suited for viral amplification in the spring, and transmission to humans in the late summer.[10] Birds remain a likely reservoir for WNv in western North America, and in particular crows, robins, magpies, and blue jays seem to be the major participants in the mosquito–bird amplification cycle.

Geographic Dimensions of West Nile Virus

The geographic analysis of disease can provide an important conceptual framework for the prediction and explanation of outbreaks. In many cases, information about disease patterns and epidemiology in endemic settings can be used to anticipate how a disease may behave in new settings. By combining information from prior disease environments with information about new or expected disease environments, researchers and policy makers are better equipped to predict and control disease outbreaks. When combined with methods in spatial statistics, the analytical tools available in geographic information systems (GIS) offer a cost-effective and timely method for integrating clinical expertise and epidemiological theory into a decision support system useful for the management of new or migrating infectious diseases.

With a few notable exceptions[12–14] most research in the WNv has ignored the geographic dimensions of infection, particularly in North America. This is somewhat surprising given the likely roles of the climate, physical environment, and biogeography on the current and future spread of the disease. The transmissibility of WNv vectors varies considerably with temperature[15,16] and features of local environments, such as the availability of standing foul water and altitude.[14,17] Seasonal survival and dispersal of nonmigratory birds may explain the geographic spread of WNv at smaller scales, particularly in the western hemisphere.[18] The migration behavior of other birds could be critical for understanding and predicting geographic "leaps" at larger scales.[19–21] Studying WNv within these broader geographical contexts not only helps to better understand the epidemiology of human illness associated with WNv, but may also contribute to a more general understanding of the spread of other insect-borne diseases, particularly under various climate change scenarios.

Spatial models are well suited to combining different sources of geographic information about disease and the environment to help understand geographic processes and relationships. Spatial modeling differs from nonspatial modeling by either controlling for or directly incorporating spatial autocorrelation within the modeling process. Spatial autocorrelation describes the tendency of observations that are close to each other in geographic space to share similar properties. For WNv in North America, spatial autocorrelation could result from the various dispersal mechanisms involving birds, mosquitoes, and humans as well as common environmental risk factors. By adopting a spatial modeling strategy, the dispersal and diffusion associated with WNv can be combined with *in situ* influences on human infection, such as demographic and socioeconomic variables,[22] to provide a more comprehensive understanding of local and regional influences on WNv in human populations. Furthermore, these approaches can smooth local stochasticity in observed data, improve model specification, and manage inferential problems associated with autocorrelated data.

To aid in the allocation of resources that mitigate future outbreaks and to better understand the geographic nature of WNv in the North American prairies, we use spatial and nonspatial models to predict municipal-level risk of human WNv infections based on a single year of laboratory confirmed human infection data. We compare how these various methods might be used to predict municipal-level incidence of future outbreaks if/when they occur. Finally, we discuss how geographic methodologies can be useful in situations where diseases migrate to new areas of the world.

APPROACH TO PREDICTING MUNICIPAL-LEVEL INFECTION

Data

Our analysis is performed at the level of the municipality, and includes 503 Alberta cities, towns, and villages (which we collectively refer to as "municipalities") that have a population greater than 100 people (FIG. 1).

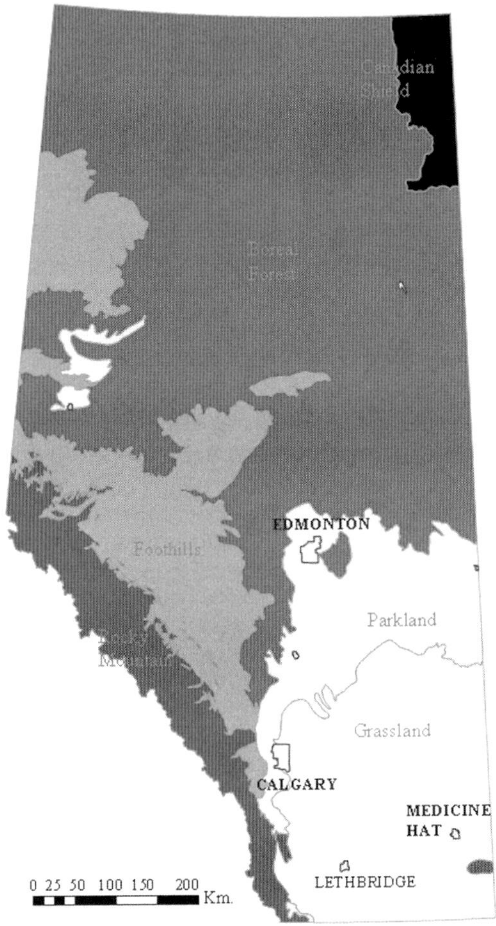

FIGURE 1. Province of Alberta, Canada, with major urban municipalities and natural ecological regions.

Several data sources are available at the municipal level. Human infection and testing data were acquired through a communicable disease-reporting system administered by the provincial health ministry. Since most persons infected with WNv are asymptomatic, our reported cases are an indirect indicator of the real presence of disease in the population. Infected bird data were acquired through a passive surveillance system in which the public turned in dead birds. We also incorporated irrigation, ecological, and urban/rural information into the model. These variables were meant to capture the environmental suitability of a municipality for sustaining a population of *Cx. tarsalis*, as well as describing the likelihood of human–vector contact. Irrigation data were included in the model as a dichotomous indicator variable. Municipalities within a 20 km distance of an irrigation district were considered within an area of irrigation. This 20 km buffer is meant to take into account the short-range dispersion characteristics of *Cx. tarsalis.*[10] We classified municipalities with a population greater than or equal to 10,000 persons as urban. The remaining municipalities are considered rural. Alberta can be subdivided into six different natural regions that represent a number of different ecological characteristics, such as physical landscape, temperature, and precipitation. The Grasslands and Parkland regions, the hottest and driest regions of the province, were entered into the model as separate dummy variables. We put the Boreal Forest, Rocky Mountain, Foothills, and Canadian Shield regions into a single indicator variable. These regions are not known to support the *Cx. tarsalis* habitat.[11]

Comparative Analysis of Spatial and Nonspatial Modeling

We combined these data into four regression models formulated to predict cases of West Nile virus infection at the municipal level.

Models A and B

Ordinary least squares (OLS) regression is a simple, popular, and powerful approach to modeling numeric data. Model A is an OLS model in which the rate of human WNv infection in a municipality is the response variable predicted by the covariates described above. Given the small number of positive WNv cases overall and the differences in population across the province, variance characteristics are not the same for all municipalities. This violates the heteroskedastic error assumption common to OLS models, such that even for a model with the proper selection of predictor variables, inferences associated with the parameter estimates are not conservative enough. We use model B, a weighted least squares (WLS) model, to take into account the nonstationary variance associated with rare diseases and inhomogeneous population distributions. Specifically, a weighted residual sum of squares is minimized where the weight is proportional to the inverse of the municipal variance in WNv infection rates.

Models C and D

For model C, counts of positive human cases were incorporated into a generalized linear model. Since human WNv is relatively rare, we assume a Poisson-distributed response variable and model a function of WNv case counts (specifically, a log-link of the mean response). This allows us to model municipal-level cases of human WNv. Because population differs between municipalities, the log of the municipal population was used as an offset variable in these models. By log transforming the population, we scaled the population variation between larger municipalities (such as Edmonton and Calgary) and smaller municipalities.

Spatial phenomena exhibit two general classes of patterns: first-order effects and second-order effects.[23] First-order effects (spatial trend) represent large-scale spatial variations, and can be thought of as the two-dimensional equivalent of increases or decreases in observations over time. Second-order effects (spatial autocorrelation) represent smaller-scale variations in which neighboring observations are in some way systematically similar to each other. This is somewhat analogous to the seasonal periodicity found in a time series. Although a clear conceptual distinction between these types of effects is often elusive in practice, their theoretical distinction is important for understanding the dimensions of disease distribution over space and time. Spatial models consider these effects in a number of ways. When trying to describe the variation in phenomena over space, modeling spatial trend and autocorrelation simultaneously can help smooth out localized and unimportant noise in to describe important (and possibly new) geographic patterns. In other cases, spatial models can be used to make up for missing covariates in a model, thereby, managing misspecification error. This is particularly true when there is residual autocorrelation unaccounted for by available covariates. In yet other instances, spatial models are used to get more precise variance estimates for model coefficients when nearby observations are known to be correlated with each other. This adjustment process is somewhat analogous to correcting variance estimates in experiments with repeated measures.

West Nile virus is not normally communicable between people; therefore, we would not expect any autocorrelation that is the result of direct communication between individual persons at the municipal level. Specifically, the frequency of human infection in one municipality should not have much direct impact on the frequency of human infection in a nearby municipality. However, autocorrelation may also arise when features that influence frequency of WNv are themselves spatially autocorrelated, such as the interaction between vectors and birds. Since the communication of the virus between the bird reservoirs, the vectors, and humans is almost certainly influenced by features of the environment, it is important that spatial models of West Nile virus take spatial autocorrelation into account when estimating or trying to understand the spatial distribution of disease.

For model C, we ignore the role of spatial autocorrelation, and assume it either does not exist, or has been accounted for by the covariates in the model. For model D, we employ a strategy that incorporates spatial autocorrelation by including the municipalities as random effects and then specifying a spatial covariance structure that defines the range and magnitude of autocorrelation. This covariance structure defines the degree to which neighboring municipalities influence each other—municipalities near each other are assumed to be more similar than municipalities far from each other. Our estimates of the magnitude and range of autocorrelation are based on previous work done in this region.[24] When estimated in conjunction with the fixed model covariates described above, this approach represents a "mixed model." Similar to model C, the response variable is a log-link function of the number of positive WNv cases in humans (with the same offset as model C), and therefore, model D is generalized linear mixed model (GLMM).[25] GLMMs are used when the theoretical rationale of mixed models (comprises fixed and random effects) is appropriate, but when the assumptions of linearity required for mixed models are violated. The GLMM can be thought of as consisting of two stages of estimation—the first is an estimation of fixed parameters (i.e., the covariates) given the random effects and the second is the estimation of the random effects.[26]

We use procedures available in the SAS system (REG for models A and B and GLIMMIX for models C and D) to solve the four models.[27] Since our goal is to use these models as tools for decision support, we do not report or directly compare parameter estimates associated with these models, but rather, report the predicted risk of human WNv infection based on the model results. Although each of the four models offers predicted values of risk for each municipality, the structure of the models makes direct comparison somewhat difficult. For A and B the predicted risk is the municipal rate of WNv in humans, and for C and D, it is the ratio of the model predicted number of cases and the expected number of cases. To simplify comparisons between the models, we rank municipal risk from highest (where rank = 1) to lowest (where rank = 503) for all four models, and qualitatively compare these ranks on maps and tables in order to identify the degree to which the results for the four models differ.

COMPARISON OF THE MODEL RESULTS

In 2003, 275 positive cases of WNv were identified in Alberta. Seroprevalence estimates suggest an asymptomatic infection rate of 3 per 1,000 in the province but a rate of 46 per 1,000 in rural southeastern Alberta.[3] The resolution and sample size of the seroprevalence survey were not sufficient to estimate the presence of disease in individuals at the municipal level.

Based on evidence of WNv found in birds, mosquitoes, and humans, the WNv season started in early July, and continued until early October (FIG. 2). Evidence of WNv appeared earlier in birds and mosquitoes than in humans, and WNv-positive humans were found later in the season than birds or mosquitoes. Nonetheless, the three indicators all peak in late August to early September, which constitute the late summer months in the region.

The ranked prediction from all four models is tabulated for the 10 municipalities with the largest count of WNv cases (TABLE 1). Although the ranking

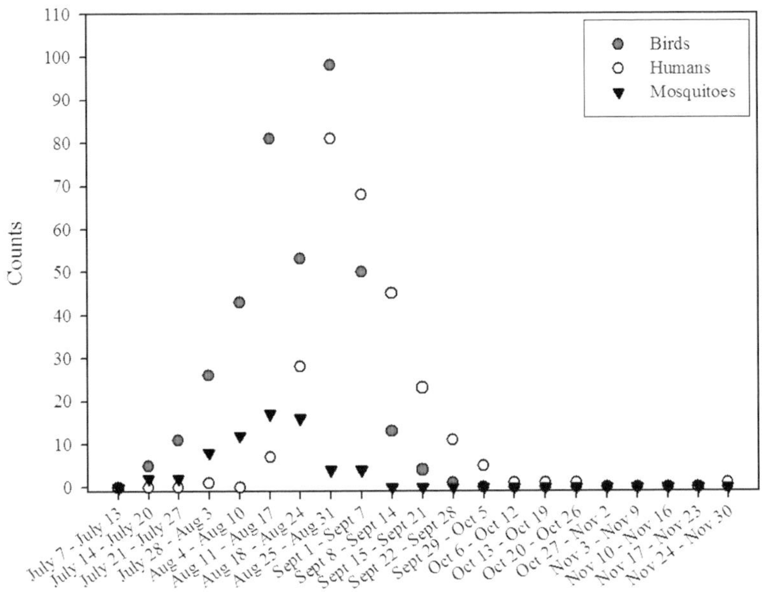

FIGURE 2. WNv in birds, mosquitoes, and humans.

is similar between all models, there is variation in the municipal ranking of human WNv risk, particularly for the larger cities (Edmonton and Calgary). Since a table comparing all 503 municipalities would have been of little informational value, we show maps with information from the four models (FIG. 3A–D). For each model, the predicted output was sorted in order of

TABLE 1. Ranking of predicted human WNv infection risk associated with municipalities.

		Modeled risk			
Municipality	Rank based on observed cases	A	B	C	D
Medicine Hat	1	27	22	19	12
Calgary	2	98	39	152	223
Bow Island	3	2	2	1	1
Edmonton	4	491	205	239	338
Taber	5	46	37	35	9
Brooks	6	48	25	112	21
Lethbridge	7	108	55	40	132
Foremost	8	3	3	4	3
Hanna	9	7	5	25	6
Redcliff	10	45	49	20	19

A = Nonspatial OLS.
B = Nonspatial OLS, weighted for heterogeneous variance.
C = Nonspatial Poisson regression.
D = Spatial Poisson regression, power covariance model.

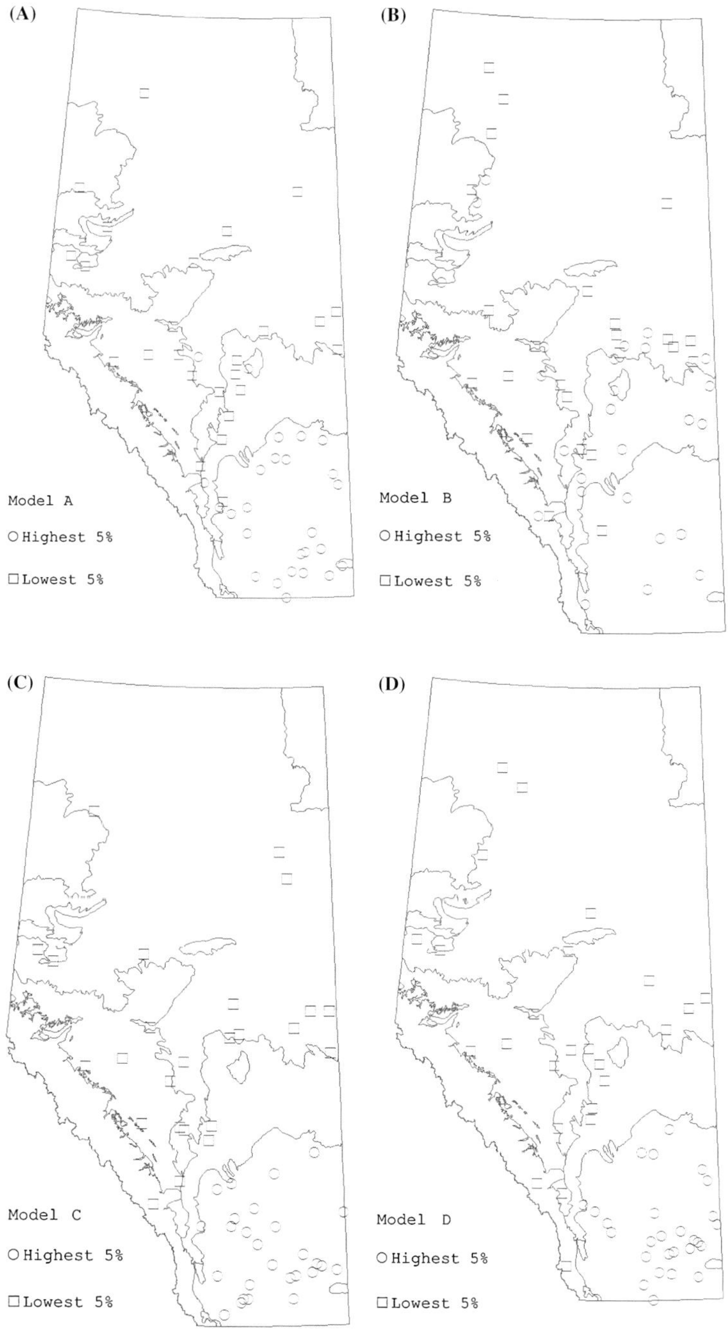

FIGURE 3. Maps of highest and lowest predicted risk of WNv infection in humans.

predicted risk of WNv associated with a particular municipality. Municipalities in the highest 5th percentile of predicted risk and lowest 5th percentile of predicted risk were mapped for all models. A qualitative comparison of the model results suggests a high level of overall similarity. For all four models, most of the highest-risk municipalities are in the southeast corner of the province, though the concentration of risk varies—with model D appearing to show the greatest concentration, and model B showing relatively more dispersal overall.

A provincial map of predicted incidence ratios (predicted cases divided by expected cases) for model D is presented in FIGURE 4. A Voronoi polygon tessellation is used to help visualize the spatial trend, with each polygon corresponding to the general location of a municipality and its surrounding area. Although the range of variation in incidence ratios is large, almost all incidence ratios above 1.00 are in the southeast region of the province. This is consistent with the observations from FIGURE 3D; most of the variation in incidence occurs in one area of the province, with the remainder of the province exhibiting very low incidence of human WNv infection. This is also consistent with provincial seroprevalence estimates of WNv.[3]

SPATIAL MODELS AND PREDICTING WNV WITH ADMINISTRATIVE HEALTH DATA

Since 2003, WNv has not been found in significant numbers in the human population of Alberta, or in western Canada in general. With only 1 year of data, it was not possible to use spatial–temporal approaches to predict future human incidence of WNv. Instead, we used a spatial model to help illustrate how risk of human infection varies across Alberta municipalities. This model smoothes out the local variability typically observed with rare diseases, and takes into account both important covariates (such as human testing data and the local and regional environment) and the structure of autocorrelation in the spatial distribution of disease. Data were obtained from a communicable disease registry and administrative health records, and the municipal-level predictions correspond well with seroprevalence estimates made in the same year. This is an important observation, since it suggests that population-based spatial modeling strategies may provide a timely and cost-effective method for estimating current incidence and future risk of infection. Undoubtedly, some dynamics of WNv are changing, and the exact character of future outbreaks may be hard to anticipate based on a single model. Nonetheless, the findings of our model are consistent with information available on the distribution of disease, and suggest that municipal-level spatial models provide a reasonable framework for analysis, surveillance, and resource allocation.

Based on evidence of WNv found in birds, mosquitoes, and humans, the WNv season appeared to start in early July, and continued until early October. Testing of birds and mosquito pools predated the outbreak period by several weeks, and therefore, we suspect the temporal trend observed is a relatively precise estimate of the WNv season. The climate on the Canadian prairies exhibits considerable variation over time and location; in the far North and the mountainous regions to the West, the short cool summers are unlikely to support a significant population of *Cx. tarsalis*. In the southwest of the province, much like the prairies across North America in general, summers are hot and long, and in general more hospitable to *Cx. tarsalis*. From year to year, the temperature can vary, triggering changes in the mosquito population, perhaps indicating earlier start dates for mosquito–bird amplification. However, given the sensitivity of *Cx. tarsalis* feeding habits to overall mosquito abundance,[17] it may be hard to use temperature or abundance of *Cx. tarsalis* as an indicator for predicting human infection.

The patterns of illness are generally similar for all four models. In spite of well-known problems associated with using OLS regression strategies to model rare diseases, model A provides similar general information as the other models, particularly when comparisons are based on rank, rather than the prediction of municipal infection rates. For model A, the highest 5th percentile of risk is located in the southeast area of the province, with the lowest 5th percentile distributed more widely. For model B, the highest and lowest percentiles of predicted risk are more spatially variable, though in general, the highest risk areas of the province are still in the southeast. For models C and D, the results are very similar; however, the cluster of highest 5th percentile appears to cover a smaller area for model D, suggesting a more localized concentration of high-risk municipalities. To some degree, the map of incidence ratios confirms this; within the grasslands region, some municipalities have particularly higher risk than other municipalities. This information could assist mosquito fogging/spraying practices in the future, helping to allocate resources to regions where the risk of infection is particularly high.

Differences in awareness of WNv, along with a higher testing rate in some municipalities, may have influenced the observed rates of confirmed illness.[24] Predicting future outbreaks is confounded by testing behavior, public concern, physician practice style, and temporal interactions between these factors. For example, a single positive case may result in public concern that results in a surge in testing, which then further increases the number of confirmed cases. It may be difficult to separate true local spikes of infection from spikes that are due to increased testing frequency until enough years of WNv infection have occurred to accumulate robust temporal data. In the future, a spatial–temporal approach to investigating outbreaks may provide information that helps untangle these issues. In the meantime, including human testing data as a covariate in disease models of human infection may help separate "noise" associated with testing variation from actual variation in infection rates.

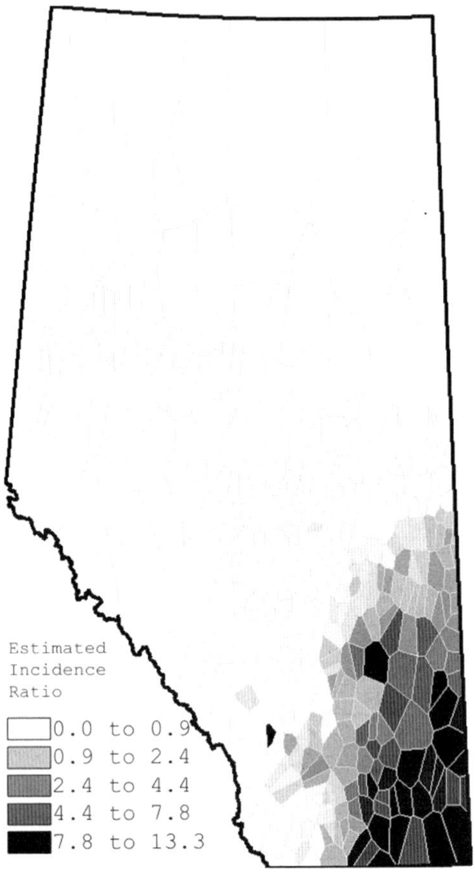

FIGURE 4. Predicted human WNv incidence ratio (based on model D).

The recent spread of WNv into new regions has resulted in an interest in how this and other infectious diseases may migrate to new locations over time. The 1999 outbreak of WNv in New York City was soon followed by a spread of infection across North America. Since the initial outbreak, human cases have been found throughout the United States and Canada, and evidence of WNv has been found in Central America. Changes in the global climate may affect patterns of severity and/or incidence of WNv infections in humans.[28] For public health planners and administrators, it is important to adopt strategies and intellectual infrastructure capable of dealing with new outbreaks in a timely manner. Recent research in WNv in North America has provided an opportunity to refine surveillance practices that could help understand and manage future changes in the distribution of other mosquito-borne diseases in North America, such as St. Louis encephalitis, dengue fever, and even malaria.

In the presence of complex spatial patterns, such as a correlation in rates of disease between geographic observations, spatial models can be used to improve the validity of statistical inferences. In a decision-support setting, and particularly at the population level, these models are important for characterizing variations in disease incidence, as well as adjusting for the inferential complications that arise from spatially autocorrelated data. When such analyses are based on administrative health data, they offer timely and inexpensive information about relative geographic variations in disease risk. With the widespread availability of routine methods to solve spatial models, and the ever-increasing quantity of electronic information in many regions of the world, these strategies offer important opportunities that support public health decision making.

REFERENCES

1. Nash, D., F. Mostashari, A. Fine, *et al.* 2001. The outbreak of West Nile virus infection in the New York City area in 1999. N. Eng. J. Med. **344:** 1807–1814.
2. Mostashari, F., M.L. Bunning, P.T. Kitsutani, *et al.* 2001. Epidemic West Nile encephalitis, New York, 1999: results of a household-based seroepidemiological survey. Lancet **358:** 261–264.
3. Ivan, M., D.P. Schopflocher, L.W. Svenson, *et al.* 2005. Estimating the infection rate of West Nile virus in Alberta. Alberta Health and Wellness. Edmonton.
4. Schellenberg, T.L., M.F. Anderson, M.A. Drebot, *et al.* 2006. Can. J. Public Health **97:** 369–373.
5. Sardellis, M.R., M.J. Turell, D.J. Dohm, *et al.* 2001. Vector competence of selected North American *Culex* and *Coquillettidia* mosquitoes for West Nile virus. Emerg. Infect. Dis. **7:** 1018–1022.
6. Hayes, E.B., N. Komar, R.X. Nasci, *et al.* 2005. Epidemiology and transmission dynamics of West Nile virus disease. Emerg. Infect. Dis. **11:** 1167–1173.
7. Peterson, L.R. & A.A. Marfin. 2002. West Nile virus: a primer for the clinician. Ann. Int. Med. **137:** E173–E179.
8. Turell, M.J., M.L. O'Guinn, D.J. Dohm, *et al.* 2001. Vector competence of North American mosquitoes (Diptera: Culicidae) for West Nile virus. J. Med. Entomol. **38:** 130–134.
9. Nasci, R.S., H.M. Savage, D.J. White, *et al.* 2001. West Nile virus in overwintering *Culex* Mosquitoes, New York City, 2000. Emerg. Infect. Dis. **7:** 1–3.
10. Turell, M.J., D.J. Dohm, M.R. Sardelis, *et al.* 2005. An update on the potential of North American mosquitoes (Diptera: Culicidae) to transmit West Nile virus. J. Med. Entomol. **42:** 57–62.
11. Darsie, R.F. & R.A. Ward. 2005. Identification and geographical distribution of the mosquitoes of North America, North of Mexico. 225–340. University Press of Florida. Gainesville.
12. Johnson, G.D., M. Eidson, K. Schmidt, *et al.* 2006. Geographic prediction of human onset of West Nile virus using dead crow clusters: an evaluation of year 2002 data in New York State. Am. J. Epidemiol. **163:** 171–180.
13. Gibbs, S.E.J. M.C. Wimberly, M. Madden, *et al.* 2006. Factors affecting the geographic distribution of West Nile virus in Georgia, USA: 2002–2004. Vector Borne Zoonotic Dis. **6:** 73–82.

14. Diuk-Wasser, M.A., H.E. Brown, T.G. Andreadis, *et al.* 2006. Modeling the spatial distribution of mosquito vectors for West Nile virus in Connecticut, USA. Vector Borne Zoonotic Dis. **6:** 283–295.
15. Dohm D.J., M.L. O'Guinn & M.J. Turell. 2002. Effect of environmental temperature on the ability of *Culex pipiens* (Diptera: Culicidae) to transmit West Nile virus. J. Med. Entomol. **39:** 221–225.
16. Reisen, W.K., Y. Fang & V.M. Martinez. 2006. Effects of temperature on the transmission of West Nile virus by *Culex tarsalis* (Diptera: Culicidae). J. Med. Entomol. **43:** 309–317.
17. Lothropm, H.D. & W.K. Reisen. 2001. Landscape affects the host-seeking patterns of Culex tarsalis (Diptera: Culicidae) in the Coachelia Valley of California. J. Med. Entomol. **38:** 325–332.
18. Rappole, J.H. & Z. Hubálek. 2003. Migratory birds and West Nile virus. J. App. Microbiol. **94:** 47S–58S.
19. Hubálek, Z. & J. Halouzka. 1999. West Nile fever—a reemerging mosquito-borne viral disease in Europe. Emerg. Infect. Dis. **5:** 643–650.
20. Rappole, J.H., S.R. Derrickson & Z. Hubálek. 2000. Migratory birds and the spread of West Nile virus in the western hemisphere. Emerg. Infect. Dis. **6:** 319–327.
21. Owen, J., F. Moore, N. Panella, *et al.* 2006. Migrating birds as dispersal vehicles for West Nile virus. Ecohealth **3:** 7–85.
22. Ruiz, M.O., C. Tedesco, T.J. McTighe, *et al.* 2004. Environmental and social determinants of human risk following a West Nile virus outbreak in the greater Chicago area, 2002. Int. J. Health Geographics **3:** 8.
23. Bailey, T.C. & A.C. Gatrell. 1995. Interactive Spatial Data Analysis. Longman Scientific & Technical. Essex, UK.
24. Yiannakoulias, N.W., D.P. Schopflocher & L.W. Svenson. 2006. Modelling geographic variations in West Nile virus. Can. J. Public Health **97:** 374–378.
25. Littell, R.C., G.A. Milliken, W.W. Stroup, *et al.* 1996. SAS System for Mixed Models. 303–320. SAS Institute. Cary, NC.
26. Waller, L.A. & C.A. Gotway. 2004. Applied Spatial Statistics for Public Health Data. 383–385. John Wiley & Sons. Hoboken, NJ.
27. SAS 9.1.3. 2006. SAS Institute. Cary, NC.
28. Paz, S. 2006. The West Nile virus outbreak in Israel (2000) from a new perspective: the regional impact of climate change. Int. J. Env. Health Res. **16:** 1–13.

Index of Contributors